SIR JOSEPH BANKS:

A GLOBAL PERSPECTIVE

SIR JOSEPH BANKS:

A GLOBAL PERSPECTIVE

Editors

R.E.R. Banks B. Elliott J.G. Hawkes
D. King-Hele G. Ll. Lucas

Production Editor
S. Dickerson

Published by
The Royal Botanic Gardens, Kew
on behalf of the co-sponsors

The conference at which these papers were originally presented was co-sponsored by the Royal Society, the Natural History Museum, the Royal Botanic Gardens, Kew, the Linnean Society of London, the Society for the History of Natural History and the Banks Archive Project.

General Editor: J.M. Lock

Typeset at the Royal Botanic Gardens, Kew, by
Pam Arnold, Christine Beard, Dominica Costello,
Margaret Newman, Pam Rosen and Helen Ward

Cover design by Media Resources, Royal Botanic Gardens, Kew

Cover portrait of Sir Joseph Banks by Joshua Reynolds reproduced by
courtesy of the National Portrait Gallery, London

ISBN 0 947643 61 3

Printed and bound in Great Britain by
Whitstable Litho Ltd., Whitstable, Kent.

CONTENTS

Poster Papers

INTRODUCTION

The papers published in this volume are edited transcripts of those presented at the conference "Sir Joseph Banks: a Global Perspective" held at The Royal Society on 22 and 23 April 1993. Motivation to organize the conference arose from the perceived need to commemorate the 250th anniversary of the birth of Sir Joseph Banks (1743–1820) by The Royal Society, The Natural History Museum, the Banks Archive Project, the Royal Botanic Gardens, Kew, the Linnean Society of London, the Royal Horticultural Society, the Royal Institution and the Society for the History of Natural History. A steering committee was set up in January 1992 and comprised:

Mr R E R Banks (The Natural History Museum) Convener
Mr H B Carter (Banks Archive Project)
Mrs S M Edwards (The Royal Society)
Dr B Elliott (Royal Horticultural Society)
Professor J G Hawkes (Linnean Society of London)
Dr D G King-Hele (The Royal Society)
Professor D M Knight (University of Durham)
Professor G Ll Lucas (Royal Botanic Gardens, Kew)
Mrs I M McCabe (Royal Institution)

The steering committee set itself the tasks of selecting a theme appropriate to cast a broad look at Banks's activities and influence worldwide, of arranging a programme consisting of topics with a balance of emphasis between the geographical and subject approach, and of inviting a group of speakers from Europe, North America and Australasia whose specialist interests and researches related to the chosen topics.

Harold Carter's biography of Banks, *Sir Joseph Banks* (British Museum (Natural History), 1988) had drawn together many of the threads of Banks's wide-ranging interests, and suggested that his influence was wider and deeper than perceived by earlier biographers. In many ways therefore Carter's biography set the scene for the penetrating views of Banks's global influence which were presented at the conference.

The conference adopted as its "logo" the portrait of Banks as a young man (1772–73) by Sir Joshua Reynolds in which the subject is seated at a desk, a terrestrial globe at his side. Over three hundred delegates attended the conference sessions and the atmosphere generated during proceedings made the illustrious setting of The Royal Society over which Banks presided for forty-two years all the more appropriate for this commemorative meeting. In addition to the papers delivered by invited speakers, there were a number of poster presentations, six of which are summarised at the end of this volume.

The resounding success of the conference and the publication of this volume could not have been achieved without financial support from the following:

The Royal Society
Royal Botanic Gardens, Kew
Linnean Society of London

Royal Horticultural Society
The Natural History Museum
Australian Government
Academic Press

In addition members of the Steering Committee express their sincere thanks to the following for their considerable help:

Dr Peter T Warren, Executive Secretary, The Royal Society
Dr John Marsden, Executive Secretary, Linnean Society of London
Julia Bruce, Banks Archive Project
Fiona Bradley, Media Resources, Royal Botanic Gardens, Kew
Suzy Dickerson, Editorial Unit, Royal Botanic Gardens, Kew
Ann McNeil, Editorial Unit, Royal Botanic Gardens, Kew

Rex E R Banks
August 1993

H.B. Carter, 'Sir Joseph Banks and the Royal Society', in *Sir Joseph Banks: a global perspective* (eds. R.E.R. Banks and others), Royal Botanic Gardens, Kew, 1994, 1–12.

1. SIR JOSEPH BANKS AND THE ROYAL SOCIETY

HAROLD B. CARTER

Yeo Bank, Congresbury, Bristol

In the historical studies of the late Georgian period of Great Britain there has been a missing element. For the historian at large Sir Joseph Banks, if he appears at all, is still effectively little more than the substance of an odd footnote. By the historian of science his place is largely overlooked because his scientific publications in the prestigious journals of his day are virtually nil. In the formal records of The Royal Society his stature, at times, seems also inversely proportional to the length of his term as President. Nor has his place in history been much helped by what has been implanted in the mind of the literate public within the past forty years as he slowly emerges from a state of neglect to gain a new image.

In 1952 a distinguished physician was the first to challenge the view of Banks as "the autocrat of the philosophers" and to declare his scattered literary relics to be "the fragments of a Colossus".(1) On the other hand in 1962 an eminent Pacific historian attached the tags which labelled Banks as "a spoilt child of fortune"; "uneducated" in the 18th century sense; a "dilettante botanist" with 'no instinct of thoroughness'; most certainly not "a man of the study".(2) And yet, in the final analysis, the same source consigns Banks to the pages of history, with pejorative acclaim, as "the great Panjandrum of British Science" — the young man who, at the age of 35, ascended "Newton's chair" to occupy it as its longest serving incumbent for nearly 42 years.

Now today, another generation on, we are bound to ask by what clouded or confused judgement did the Council of The Royal Society propose and the Fellows duly elect such a man to preside over its affairs for so long? In retrospect, perhaps, it has as much to say about The Royal Society and the state of science during his life and times as it has about this enigmatic man himself. How, indeed, did he achieve this pinnacle in the intellectual life of his period? Beyond this, also, how did he as an independent Georgian landed gentleman of England gain a form of global influence unusual at any time in modern history — certainly for a private individual — even as a President of the Royal Society.

Although born in London on 13 February 1743 at 30 Argyll Street, St James's (then), he was of Derbyshire and Yorkshire descent, the middle-class grandson of another Joseph Banks who was himself a still earlier form of F.R.S. (1730).(3) The estates of his inheritance were mostly in Lincolnshire, the county with which his name is now usually associated, but they were derived from the speculative energies of his Yorkshire-born great-grandfather from Ribblesdale, a shrewd Sheffield lawyer with a keen sense of business and property values. Successively tutored by the fens at Revesby Abbey, prepared at Harrow, disciplined at Eton and liberated at Christ Church, Oxford, Banks was the first of his family to enter a university with serious intent — apart from his step-uncle

Collingwood who died as an undergraduate at Christ Church and was buried there. As a gentleman-commoner, by the time he came of age in 1764, he had matured into the semblance of a serious student of natural history. With a questing and essentially scientific curiosity, focussed mainly on botany and entomology, these were to him "those years of education that are the happyest that life affords" — but they were largely years of self-education[4] in the relaxed and rather silent term of Dr. Humphrey Sibthorp as the incumbent Sherardian Professor of Botany. He came down from Oxford at the end of summer in 1764 without any formal degree as an early English exponent of the new Linnaean system of classifying the natural world, though by no means uncritical of its tenets. To these first insights he now added the intellectual stimulus to be found at the British Museum in its first decade among the small group gathered in the Department of Natural and Artificial Productions round Dr. Daniel Solander, a Linnaean apostle direct from the Uppsala fountainhead itself.[5]

Within a year, at the ripe age of 23, in February 1766 he was nominated as a Fellow of The Royal Society, though not yet noted for any work remarkable "in a philosophic way" but, in the words of his five sponsors on his election certificate as "likely (if Chosen) to prove a valuable member". In other words, he was a young man of whom already there were great expectations.[6] Then, in May

The Meeting Room of The Royal Society in Crane Court, 1710–1779, where Banks was elected a Fellow, 1766, and President, 1778.
Reproduced courtesy of the President and Council of the Royal Society of London.

The Meeting Room of The Royal Society in Somerset House, 1820–1849, where Banks sat as President, 1780–1820.
Reproduced courtesy of the President and Council of the Royal Society of London.

1766, he was elected a Fellow "as any decent tradesman might then have been" according to Barrow, glancing down from the superior viewpoint of the mid-19th century.[7] But Banks, at his election, was at sea on his first voyage abroad, not "as any decent tradesman might then have been" but as a sort of post-graduate biologist, an accepted supernumerary on a 32-gun frigate of the Royal Navy on fisheries patrol off the coasts of Labrador and Newfoundland. Here Banks spent a mosquito-ridden summer as an active field naturalist toughened by his first sobering bout of malaria in its relatively benign tertian form but relieved after several weeks by his recourse to "the bark", a decoction of Peruvian cinchona.[8] He returned to London in January 1767 as an experienced collector with the first properly documented specimens in natural history from that region, as emphatic evidence of his scientific merit as a young F.R.S. He signed the book at Crane Court on 12 February 1767, two weeks after his return and one day before his 24th birthday.

Almost an exact year later he dined at the Club of the Royal Philosophers in Middle Temple Hall as the guest of the President of The Royal Society on 4 February 1768 and again on 11 February as the guest of the Treasurer. These were critical occasions during the preparation of the Society's "Memorial" to the King for a grant of £4,000 towards a Royal Society expedition to the South Seas to observe the transit of Venus in June 1769. From these two convivial occasions in

Middle Temple Hall may be dated his commitment to the voyage of HMS *Endeavour*, on his own initiative and at his own expense with the blessing of The Royal Society Council — three months before the appointment of James Cook as its Commander.[9]

In the spring of 1767 he had acquired his first London house at No. 14 New Burlington Street as a haven for his growing library and natural history collections and their first graphic recording by Georg Ehret and Sydney Parkinson. On 16 August 1768 he left this house to join HMS *Endeavour* at Plymouth. On 13 July 1771 he returned to the same house with the first natural history and ethnological collections to be seen in Great Britain from the South Pacific, unassailable proof that his Fellowship and the trust of the Council of The Royal Society had not been conferred in vain. As Fellows of the Society Mr. Joseph Banks and Dr. Daniel Solander had carried the biological sciences into the South Pacific for the first time and had opened new fields of scientific enquiry in exploration for the inquisitive white European.

In a rumpus of mutual misunderstanding with the Admiralty he withdrew from the second Cook voyage to the Pacific. Instead, his own last exploratory voyage abroad to Iceland in 1772, improvised under his own command and, again, at his own expense, was in its lesser way well done as scientific field work. It set the northern limit to Banks's own personal collecting sites whilst Tierra del Fuego had marked the southern margin — a global range in latitude no naturalist before him had achieved. At 29 in his chosen scientific field he had done more than most Fellows of the Royal Society "for the Promotion of Natural Knowledge" as an intellectual entrepreneur of his time. This was before the days of well-devised university courses, Parliamentary grants-in-aid, or wealthy foundations springing from industrial success, creating colleges and sponsoring expeditions to unexplored stretches of Planet Earth.

The Royal Society expedition of HMS *Endeavour* 1768–71 stands as a striking partnership between two remarkable men — Joseph Banks at 25 and James Cook at 40 — broadly harmonious in their working relations on a small and crowded ship in a three-year circumnavigation through strange seas. It is unique among British voyages of exploration. No other was conceived or conducted on such a plan, as a project in physical science funded from the Royal purse combined with one in the biological sciences funded from a county rent roll, executed under Admiralty management. All these elements were compounded without undue heat of fusion into a pioneering venture which was broadly successful in its main intentions, though scarred at the end by heavy human loss.

For Banks himself, by the end of 1772, his three "voyages of curiosity" in pursuit of "Natural knowledge" were the undoubted foundation of his future as a dedicated servant of Science — comprehensive in their vision; thorough in their preparation; professional in their collections; detailed in their documentation; and well conserved at the end for the scientific world to study even to the present day in the case of the botanical material. These were not the operations of a "dilettante botanist" with "no instinct of thoroughness". They were the far-sighted enterprises of an intellectual man of action — one who was, in fact, a man of the study as well as of the field.

Within seven years of his return from the South Pacific, by 1778, Banks had created at No. 32 Soho Square a well-organised "Academy of Natural History" preeminent in its botanical specimens and the conditions for their study and backed by a library unique for its day in the range of its coverage in natural history.[10] The death of Daniel Solander and the economic stringencies after the American War of Independence broke any prospect of co-authorship on the South Pacific flora to the distress and confusion of botanists ever since. But, instead of scientific publication, a world reference centre had been established in Soho Square for the next half century, enhancing the list of plant species published in the *Species plantarum* 1762–63 of Linnaeus by about one-fifth from a single expedition based on private capital and on which the international scientific world could draw more freely during years of global unrest than from any other single national establishment then at hand.

When Banks was elected as President of The Royal Society on Saint Andrew's Day 1778 he had been a truly active presence in its affairs for nearly seven years. He was a member of Council for three years, 1773–74, 1774–75, 1777–78, in the diverse company of such men as Henry Cavendish, Benjamin Franklin, Dr. Nevil Maskelyne, Dr. (later Sir) William Watson, the Hon. Daines Barrington, Dr. Daniel Solander, Captain the Hon. Constantine John Phipps (later the 2nd Baron Mulgrave), John Hunter, Major William Roy and Peter Woulfe. He was, indeed, a member of that Council which proposed the general plan for Cook's third and last voyage, the published account of which Banks would guide through the press, during 1780–84, in the first somewhat uneasy years of his Presidency. At all times in these seven years he attended assiduously the regular meetings of the Society but not always as a diner with the Club of the Royal Philosophers. For two years, 1775–76 and 1776–77, he was off the Council apparently at his own request, and on Thursdays dined instead as the elected "Perpetual Dictator" of his own "rebellious" Royal Society Club.[11] Its members were nearly all his seniors by some years and not a botanist among them. All were inclined to the non-biological sciences — physics, chemistry, survey, navigation, or the management of large estates. It is on the whole very clear that, in the years before his election as President, Banks had imposed on a wide swathe of the Fellowship not only his strong personality but the force of his intellect and range of scientific understanding. Supporting this was the evidence of his organising power in the record of his voyages and in the conservation and scientific use of his collections and library at 32 Soho Square.

In the Presidential election at Crane Court in 1778 Banks succeeded by a handsome majority of some 220 votes, although the retiring President, Sir John Pringle, apparently preferred Alexander Aubert (FRS 1772) the London merchant and excellent amateur astronomer, an older man by 13 years. That evening, however, the seed of future disturbance to the peace of the Society was also sown by the election, also by a large majority, of the Reverend Paul Maty (FRS 1771) as the Principal Secretary to succeed the Reverend Dr. Samuel Horsley (FRS 1767). Trinity, Cambridge as Secretary to Christ Church, Oxford, in the Chair was an uneasy pairing fore-shadowing the tensions yet to come.

After his marriage on 23 March 1779 and his leasing of the villa with its surrounding acres at Spring Grove near Hounslow in the same year the life of Banks settled into a regular pattern thereafter determined by the seasonal operations of The Royal Society. From the first week in November he was at hand in London, either at No. 32 Soho Square or, after 1780, at Somerset House in the Strand on the Society's business, preparing for Saint Andrew's Day and its crucial elections. Each year from then on he was assiduous in his attendance at

Somerset House on Thursday and Friday each week for Council meetings and the presentation of papers of which he himself read or otherwise supported many of those accepted. In his long tenure as President he attended well over 90 per cent of all such occasions, seldom calling on one of his Vice-Presidents to take his place and that mostly in his last decade on grounds of extreme ill-health. This implies also that he was equally regular in the chair at the dinners each Thursday evening of what from his time became known as the Royal Society Club.(12) These were held at the *Crown and Anchor*, just off the Strand by St. Clement's Dane, and preceded the Society meetings which began at 8.00 pm in Somerset House nearby. His diligence as a dining Clubman was exceeded only by that of Henry Cavendish, his close friend, and a constant visitor at No. 32 Soho Square for Banks's own weekly *soirées* and other more casual occasions.

The formal working year of the Society ended usually in late June or early July. Then from 1779, in about mid- to late August each year the Banks family — Sir Joseph, his wife Dorothea and his sister Sarah Sophia — set off north to Revesby Abbey in the heavy family coach with a hired coachman, postillion and four hired post horses. This was a 3-day direct journey but for some twenty years, 1793–1812, the route lay via Overton Hall, near Ashover in Derbyshire, adding another 2–3 days on the road. Thus, during September and October Banks dealt with his estate and county business, but at all times, was receptive also to that of The Royal Society. This was mainly concerned with the printing of the *Philosophical Transactions* and those papers accepted for publication but it also involved problems in the choice of papers for reading in the sessions to come. The smooth flow of this correspondence, Royal Society or otherwise, was ensured by the diligent presence of Jonas Dryander at Soho Square, which was effectively Banks's head office and so recognised by the world at large. A letter sent from here to Revesby Abbey or Overton Hall usually brought a reply to Dryander in five days. Indeed, apart from his duties as Royal Society Librarian for 25 years, Dryander in this way was also an effective Assistant Secretary to the Society for rather longer — more nearly 30 years.

During the 40-odd years of Banks's term as President, of the papers accepted for reading nearly 1,000 were printed in the 42 volumes of the *Philosophical Transactions* from more than 300 contributors.(13) An inspection of these volumes does not support the notion that Banks was unduly biassed against mathematics, astronomy and the physical sciences. Of those published between 1780 and 1820, some 580 or 60 per cent would now be classed within the range of these subjects. Over all some 72 per cent are patently non-biological. Of the remainder that are broadly biological in substance less than 8 per cent belong to the fields of most direct concern to Banks himself — those of botany and zoology. However "sick with the lust of domination" in the eyes of his most rabid critics(14), Banks certainly did not override the judgement of that part of the Council which formed the Committee of Publications to the detriment of mathematics and the physical sciences. Indeed, during the very years when Peter Pindar was so clamorously vilifying Banks as the enemy of those sciences he was heavily engaged in his role as PRS in fostering their advance.

Here, perhaps, there is no more important element than his long and close friendship with the Hon. Henry Cavendish and the various scientists who looked to Cavendish as a shining star in the experimental sciences of physics and chemistry. It was Cavendish on 16 April 1784 — when The Royal Society "dissensions" were at their aggravating climax — who walked the intended course of the baseline on Hounslow Heath with Major William Roy and Banks to set the survey in motion.[15] From their first years on the Council together to the end of his life he remained a constant figure with Banks at the Club dinners and the Soho Square evenings. It is probably no coincidence either that Cavendish's library was for so many years in Dean Street not far from the back entrance to No. 32 Soho Square; nor that Cavendish was from time to time a guest of Banks at Overton Hall when matters of engineering concern were discussed with Matthew Boulton.[16] More important, perhaps, is it to note that during the troubles at The Royal Society in 1783–84, when the place of Banks as President was seriously threatened, it was the sane advice of Cavendish and his steady support of Banks, probably more than any other single factor, that ensured the return of peace to the Society.

In relation to this episode there still remains the legacy of the probing pen of the satirist in prose or verse or of the cartoonist in the coloured prints, to season, and in some measure corrupt, the records that survive. Peter Pindar, as the widely read tabloid press of his day, harped endlessly on the life-style and personal habits of the King and the Royal family. When that ceased to engage the public mind he turned on Banks as a friend of the King and as a mere bug "in the sacred chair where Newton sat", enshrined in his fractured verse derived from the gossip and innuendo of dissident elements in the Royal Society. Banks was impaled as a sorry contrast to the men

"... whom Science so reveres
HORSELY and WILSON, MASKELYNE, MASERES,
LANDEN, and HORNSBY, ATWOOD, GLENIE, HUTTON —"

and so on. In this way Pindar preserved for posterity "Some Members of the Minority" the hard core of those who stimulated the "dissensions" in The Royal Society which effervesced in the winter and spring of 1783/84.[17] A small Cambridge-led group of mathematical practitioners of varying merit in their field, they contested the management changes in the Society emerging from the, as yet unpracticed, energies of a young President from Oxford imbued with the biological sciences in their early form. That they failed to unseat Banks, in spite of their pamphlet campaign, owed much to the growing appreciation in the Society that, though Banks was clearly much less than a Newton, he was certainly a very great deal more than a Macaroni or a "dilettante botanist". The weight of opinion mobilised under the influence of Henry Cavendish was almost certainly decisive.

Not since John Evelyn and the foundation of The Royal Society under Charles II had the Society enjoyed such a close relationship with the reigning monarch as that established by Banks with George III. From their first meetings at St. James's Palace and the White House, Kew, in August 1771, it was an active friendship of forty years, stifled only by the last prolonged illness of the King and the advent of the Regency. The first evidence of this symbiosis was the despatch of Francis Masson to the Cape of Good Hope in 1772 as the first of the long line of Kew collectors. Though nominally this was in response to an application by The Royal Society in the name of its President Sir John Pringle it sprang from those first discussions between the King and Banks on the botanical harvest gathered on the *Endeavour* voyage.[18] All that evolved thereafter in the development of the Royal Botanic Garden, Kew, and the Spanish Merino stud at

Marsh Gate, Richmond, derived from the personal relations between the King and Banks as friends and colleagues. By contrast, in the management of the baseline survey on Hounslow Heath by William Roy FRS it was Banks in his role as PRS who dispensed and accounted for the funds received for this from the Royal purse. In this pioneering operation he supervised the manufacture and production of the precision instruments needed for the triangulations that ensued, 1783–90 — including his 3-year goading of Jesse Ramsden FRS to produce "the great 3-foot theodolite of the Royal Society" that served the Ordnance Survey until 1853.[19] Concurrent with these terrestrial matters was the King's recognition of William Herschel (FRS 1781) after his introduction by Banks, following the discovery of the "Georgium Sidus" later accepted as the eighth planet Uranus. Advised by Banks, the King provided Herschel with a generous pension and subsequent large grants toward the enlargement and improvement of his telescopes to reach their peak in the giant reflector and 40-foot telescope of 1795. In all these diligent explorations of outer space Banks served again as accountant to the King for the royal disbursements.[20]

On scientific matters Banks was the voice of The Royal Society enlightening and advising the King, mostly in their friendly walks by the Thames at Kew and Richmond, their occasional rides together through the Great Park, in conversation on the south terrace at Windsor, but also more formally in his written advice or direction to the King's ministers or officers of the civil service. Informally he was the supremely knowledgeable man of affairs translating the everyday world to his Royal master but in his personal relations with the King he was silent on current matters of policy or political debate, subjects on which he felt the King himself was sometimes too free in their conversations together. Adamant in his refusal to seek Royal favours either for himself or others the only honour he accepted was admittance to the Order of the Bath in 1795. As the first civilian recipient of this honour he accepted it only for its relation to his place as President of The Royal Society, free from political or personal elements. As such it stands as a bench mark in the rising status of science in the national scale of values and as such Banks wore the red ribbon in the chair at Somerset House thereafter and on other appropriate public occasions.

Among the predecessors of Banks as President of The Royal Society Sir Hans Sloane (FRS 1684, PRS 1727–41) stands preeminent as the biologically-inclined virtuoso. Banks himself marks the transition from such 18th century virtuosity to 19th century science. The subsequent trail is well marked by a succession of distinguished Fellows of the Society all elected after their returns from Southern Hemisphere voyages in the wake of Banks and his self-financed and individually mounted part in that pioneer Royal Society expedition to the South Seas in HMS *Endeavour*. First, in the new century, is Robert Brown (FRS 1811) of the tragic but productive voyage of HMS *Investigator*, opening new vistas of botanical science on the foundations laid by Banks and Solander 30 years before. Brown forms a strong link with Charles Darwin (FRS 1839) and the voyage of HMS *Beagle*, extending the range of Southern Hemisphere collectors and the intellectual boundaries of the biological sciences beyond all question. Then come Banks's successors as Presidents of The Royal Society: Joseph Hooker (FRS 1847, PRS 1873–78), preeminent as a botanist, building a career from his voyage on HMS *Erebus* along Southern Hemisphere tracks known first to Banks; and Thomas Huxley (FRS 1851, PRS 1883–85) in HMS *Rattlesnake* following his zoological bent on the Australian east coast and Barrier Reef well known to Banks. Then, at the end, a little more than 100 years after the voyage of HMS *Endeavour* comes that of HMS *Challenger* with Henry Moseley (FRS 1877) on board to expand the zoological horizons in marine biology first glimpsed by Banks in the South Sea with his dipping net from his small boat adventures in "the great pacific ocean" when the sea was calm enough.(21) All these distinguished Fellows of The Royal Society stand as evidence of the far-reaching influence of Banks in time and space on the evolution of the biological sciences as we perceive them now. His scientific place in his own time is attested only by his long tenure as President of the Society and the material evidence of his collections, manuscripts and library — not by any impressive swathe of scientific tomes and papers but rather by the extensive use of his sources made by others, acknowledged or not as time has shown.

As the Phillips portraits of 1808–10 so clearly show, Banks was certainly an authoritative figure in the chair at Somerset House, though we must remember that these were painted of a man often sweating in the agonies of gout before he found the palliative virtues of colchicine as the active principle of Husson's *eau mêdecinale*, thanks to Lady Spencer in 1810.(22) But, in the Presidential chair or not, his stoic endurance of pain, as Hatchett has observed,(23) enabled him still

> "... to converse calmly on subjects of Science and business when it was evident ... that he was suffering excruciating torture ..."

Charles Hatchett (FRS 1797) knew Banks well during the last 25 years of his life in the chair, in meetings of the Council and as a colleague grappling with scientific problems. His view of the man therefore is well worth noting here: (24)

> "Sir Joseph Banks without pretending to be deeply versed in any branch of Science possessed nevertheless no small share of scientific knowledge, and moreover he was a most useful sort of living Index on all occasions when accurate and profound information was required upon any subject connected with the various Sciences ..."

This is not exactly the image left by Humphry Davy of a man "without much reading and no profound information".(25) Nor is it that of someone who was "uneducated in the 18th century sense" and "not a man of the study".(26) If to this we further add Hatchett's observation, written as someone who often argued strongly with Banks, that (27)

> "... although he might previously have formed some particular opinion yet he was commonly open to conviction and did not pertinaciously adhere to it ..."

this modifies the image of an "autocrat of the philosophers". As a contemporary judgement this is confirmed also by the nature of the man emerging from the substance of his wide-ranging correspondence and other papers at last gathering in a coherent assembly now.

Let me conclude with the question — How much was the global repute of Banks in his lifetime derived from his long tenure as President of The Royal Society? Certainly The Royal Society was the axis of his manifold activities, and a service to the rise of science in which he took great pride, but there was apart from this his personal eminence as the sage of Soho Square. Thunberg in 1778 had rated No. 32 as "an Academy of Natural History" and so it truly was by the standards of its time.[28] However, within it there was another element of which during his life few if any knew its magnitude, epitomised in the manuscript files gathered in the small study especially after the younger Pitt's accession to power in 1783. During the ten years of relative peace until the fatal year of 1793 the Government turned increasingly to Banks as the boundaries of British trade expanded into the relative vacuum of the vast Pacific region. In this context a noted Commonwealth historian in 1964 at last draws Banks from the shadow of a footnote into the clear light of the text as the "backroom boy" who

> "... combined the persistent curiosity of the scientist with a merchant's practical exploitation of opportunities ..."

in;his advice to those engaged in building what we now view as the second British empire.[29] For a decade Banks was *de facto* a Privy Councillor. From March 1797 he was a sworn and active member with all that this implied in the affairs of the nation then and in his relations with the King. His little study in Soho Square became effectively another office of "the great Council" with its own weight and influence separate and distinct from his place in the chair at Somerset House. At the Privy Council committee meetings in Whitehall Banks sat as the widely travelled knowledgeable man of affairs of notable integrity on whom successive Governments could rely, from the first administration of the young Pitt to that of the 2nd Earl of Liverpool.

Nonetheless, as the President of The Royal Society for so long — apart from his presence *ex officio* on various public bodies, such as the British Museum and the Board of Longitude — he developed a network of international scientific communication sustained through more than twenty years of global war. In this he carried the influence of The Royal Society far beyond the limits of national academic rivalries as far as this could then be done. Neatly defined by William Eden as His Majesty's "Ministre des Affaires Philosophiques" [30] he became, in more modern terms, for 30 years or more in effect, a sort of Permanent Secretary to a Ministry of Science and Technology whose services were at no time a burden on the Treasury either for salary or pension.[31] But, whatever the pigeonhole of history in which he may at length be fitted, there is no doubt that, in his term of office as President, The Royal Society grew in stature as an advisory body to Government and that Banks represented it as an honest but tough and resourceful negotiator of British scientific interests, especially during the wars with France. In this role he stands high as a great figure in international science. In the history of The Royal Society, however, he must find his place as a bright luminary of the Georgian enlightment of Great Britain, spanning the transition in the pursuit of natural knowledge from the 18th century intellectual virtuoso to the professional scientist as a Fellow of the Society we recognise today.

Notes

(1) H.C. Cameron, *Sir Joseph Banks, K.B., P.R.S.: the autocrat of the philosophers*, London, 1952, 280.

(2) J.C. Beaglehole (ed.), *The Endeavour Journal of Joseph Banks* (*1768–1771*), Sydney, 1962, 23–4, 123–4.

(3) J.W.F. Hill (ed.), *The Letters and Papers of the Banks Family of Revesby Abbey 1704–1760*, Hereford, 1952.

(4) J. Banks, 1768, 'Journal of an excursion to Wales and the Midlands began August 13th 1767 ending January 29th 1768', University Library, Cambridge, MS.6294.

(5) H.B. Carter, *Sir Joseph Banks 1743–1820*, London, 1988, 27.

(6) Ibid., 31–2

(7) J. Barrow, *Sketches of the Royal Society and Royal Society Club*, London, 1849, 39.

(8) A.M. Lysaght, *Joseph Banks in Newfoundland and Labrador, 1766: His Diary, Manuscripts and Collections*, London, 1971.

(9) Carter, op. cit. (5), 59.

(10) C.P. Thunberg, *Travels in Europe, Africa and Asia, made between the years 1770 and 1779*, London, 1795.

(11) T.E. Allibone, *The Royal Society and its Dining Club*, Oxford, 1976.

(12) A. Geikie, *The Annals of the Royal Society Dining Club. The Record of a London Dining Club in the Eighteenth and Nineteenth Centuries*, London, 1917.

(13) Carter, op. cit. (5), appendix XIV, XV, XVI.

(14) Some Members of the Minority, *An History of the Exclusions from the Royal Society, which were not suffered in the course of the late debate* ... London, 1784.

(15) Carter, op. cit. (5), 202–5.

(16) Ibid., 345.

(17) J. Wolcot, *Peter's Prophecy; or The President and the Poet, or An Important Epistle to Sir J. Banks, on the Approaching Election of a President of the Royal Society*, London, 1788.

(18) J. Banks, 1782, Wellcome Institute for the History of Medicine, MS.5219.

(19) Carter, op. cit. (5), 207–9.

(20) Ibid., 178–180.

(21) Ibid., 88, appendix II.

(22) Ibid., 530–1.

(23) C. Hatchett, 'Banksiana', private MS, f.1.

(24) Ibid., f. 14.

(25) E. Smith, *The Life of Sir Joseph Banks, President of the Royal Society with some notices of his friends and contemporaries*, London, 1911, 300.

(26) Beaglehole, op. cit. (2).

(27) Hatchett, [1830], op. cit. (23), f. 14.

(28) Thunberg, op. cit. (10).

(29) V.T. Harlow, *The Founding of the Second British Empire.* London, 1964.

(30) Smith, op. cit. (25), 207.

(31) H.B. Carter, 'Sir Joseph Banks — the cryptic Georgian'. *Lincolnshire History and Archaeology*, (1981) **16**, 53–62.

M. Crosland, 'Anglo-Continental Scientific Relations, c. 1780–c. 1820, with special reference to the correspondence of Sir Joseph Banks', in *Sir Joseph Banks: a global perspective* (eds. R.E.R. Banks and others), Royal Botanic Gardens, Kew, 1994, 13–22.

2. ANGLO-CONTINENTAL SCIENTIFIC RELATIONS, c. 1780–c. 1820, WITH SPECIAL REFERENCE TO THE CORRESPONDENCE OF SIR JOSEPH BANKS

MAURICE CROSLAND

University of Kent, Canterbury

Science depends very much on co-operation, not only between people in the same country but internationally. In the late eighteenth century scientific advance was concentrated very much in western Europe.(1) In this chapter it is appropriate to focus on the period of Sir Joseph Banks as an international figure through his presidency of the Royal Society (1778–1820). In political history the period chosen includes nearly a whole decade of the *ancien regime* before the coming of the French revolution.

In science the period was one of great activity in chemistry, especially by Lavoisier, who was building up a whole new interpretation of the subject which culminated in what is usually described as the "chemical revolution". Lavoisier drew extensively on the work of others, notably the British pneumatic chemists including Priestley, who actually visited Paris in October 1774 and told Lavoisier about his preparation of a gas, later to be called oxygen. More important still was a visit by Charles Blagden to Paris in June 1783. Blagden told Lavoisier that Henry Cavendish had recently tried to burn "inflammable air" (hydrogen) in closed vessels and had obtained a small amount of water.(2) Cavendish had not taken this any further, but for Lavoisier this was a vital piece of information. He repeated the experiment and used it to show that water is not a simple substance but a compound of what he would later call hydrogen and oxygen. This work was vital in reorganising chemistry around the oxygen theory to replace the old phlogiston. Another visitor to France before the Revolution was the agriculturist Arthur Young, who undertook three journeys in that county in the years 1787–90.(3) In 1787 he met Lavoisier and visited his laboratory, which he later described for the benefit of his British readers.

There were always more British visitors to the continent, and particularly to Paris, than continental visitors to Britain. Yet one feature of Britain that attracted visitors was the tremendous growth of certain branches of industry, a development often described as "the industrial revolution". Some visitors wanted detailed information of the latest British inventions so that these could be copied and used back at home.(4) Watt's improvements of the steam engine and the mechanisation of the cotton industry attracted special attention. Yet since this involved technology rather than science, and the two were largely distinct at this time,(5) it will not be necessary to pursue this subject further, except to say that one of the effects of the revolutionary and Napoleonic wars was to cut off the continent from access to many important technical innovations. Thus by 1815 the British industrial lead (except in chemistry) was greater than ever.

The French Revolution of 1789 and the war beginning in 1792/93 comes at the heart of our period. If we add the period of power of Napoleon Bonaparte, this takes us up to his final military defeat at Waterloo in 1815 and the resumption of a period of normality. The war, therefore, cut off Britain from continental travel for a period of more than 20 years. It effectively put an end to the tradition of the "grand tour", by which wealthy members of the English gentry and nobility gained first-hand knowledge of the continent of Europe. Correspondence did not cease but its frequency was greatly reduced.

France and Britain were the two great powers of the late eighteenth century. They were also the two countries who possibly contributed most to science in this period. Add to this the fact that France is Britain's nearest continental neighbour and the traditional starting off point of any continental excursion and you have several reasons why France will loom large in this paper. The fact that the war was with France was particularly unfortunate for scientific communication and was made worse by the aim of the French government from the beginning to exclude British goods from the continental market. Under Bonaparte the so-called "Continental system", establishing a severe sea blockade of continental ports, was extended.[6] Yet the system was undermined by smuggling and by the passage of goods through north German and Scandinavian ports. Also there was no intention of stopping scientific correspondence or books, and travellers on the occasional ship which crossed the Channel were usually persuaded to carry such material.

Because of the blockade, one non-French periodical deserves special mention in maintaining scientific communication between Britain and the Continent. This was the *Bibliothèque britannique*, published in Geneva by three Swiss men of science led by Marc-Auguste Pictet.[7] With agents in London, and using alternative northern routes, Pictet was able to publish monthly reports of developments in British literature, science, medicine and agriculture over the period 1796–1815. The *Bibliothèque britannique* was actually subdivided into several series, of which one was devoted entirely to science and technology. It was so successful in obtaining early news from Britain that often the Paris scientific periodicals made use of the Genevan publication for texts of memoirs in *Phil. Trans.*, etc., even though the distance travelled by the material was several times that of the direct route.

The *Bibliothèque britannique* reported on a whole range of different sciences including the work of the astronomer William Herschel. Under the heading of *la physique* there were many articles on heat, a subject of special interest to Pictet, including work by Rumford and Leslie. Pictet's personal interests may account for the almost complete absence of botany but there were many articles on the emerging science of geology. The journal contained a review of James Hutton's *Theory of the Earth* (1795) followed by a series of papers supporting Hutton's view by Sir James Hall. Thus, despite some vacillations on the part of the editor, the *Bibliothèque britannique* contributed to the rise of the Vulcanist theory of geology on the continent and the demise of the rival Neptunist school of Werner. In the field of chemistry the journal had good relations with Humphry Davy and was usually the first continental journal to publish his important Bakerian Lectures in translation.

We turn to Sir Joseph Banks, who became a central figure in British science in this period. His correspondence reveals that the majority of his contacts on the continent of Europe was with France. Indeed, his correspondence with France is greater than with all other European countries combined. Banks's formal French contacts extend back to 1772, when he was elected as a corresponding member of the Paris Academy of Sciences. The system employed under the *ancien regime* was to elect foreign members, each as the particular correspondent of a named French Academician. In the case of Banks the Academician named was the astronomer Lalande (1732–1807). Surviving letters in Britain(8) do not suggest that the correspondence was particularly intense and the correspondence the Paris archives is not much richer.(9) Nevertheless it established a valuable link between the young English botanist and French *savants*, which grew in importance after Banks's election in 1778 as President of the Royal Society. In 1787 Banks was promoted from correspondent to *associé étranger*, one of eight such places for foreign scientists of distinction. The Paris Academy was closed down in the most violent period of the French Revolution to re-emerge as the First Class of the National Institute in 1795.(10) When in December 1801 the Institute decided to elect a few distinguished *associés étrangers*, Banks was the first man of science to be elected.(11)

Just as many British people wrote to Banks in his capacity as President of the Royal Society asking to become members, so did many *savants* and *demi-savants* from the continent of Europe. They included the astronomer Jeaurat (1724–1803), the traveller and explorer Sonnerat (1749–1814) and the botanist Antoine Laurent de Jussieu (1748–1836). It is interesting to note that, whereas British members might be elected on no stronger grounds than an expression of an interest in natural philosophy,(12) after the early years of the society it was much more difficult for a foreign member to be elected. In 1761, in an attempt to limit the number of foreign members, the Royal Society stipulated that foreign candidates must have their certificate signed by at least three Foreign Fellows as well as three British Fellows.(13) Many foreign candidates therefore never reached the stage of a formal election. Although Jussieu was considered in 1788, it seems that there was some antipathy to his botanical theories and he was rejected.(14) Incredibly he had to wait another forty years to become F.R.S.(15) Lavoisier only had to wait a few months. There is a letter from Lavoisier to Banks dated November 1787 saying that he was honoured to be considered.(16) He was one of a number of Foreign Fellows to be elected in April 1788.

A recent study of the Banks correspondence has shown that the overwhelming majority of letters he received were in the period of 1781 to 1805, with a first peak in 1786 and a second peak in 1801, corresponding to the Peace of Amiens.(17) One of the most assiduous of Banks's French correspondents was the botanist, P.M.A. Broussonet (1761–1807), who was successively a member of the Royal Academy of Sciences (1785) and subsequently of the First Class of the Institute. Broussonet wrote Banks more than a hundred letters from Paris between 1782 and 1791, together with some later letters from Spain and Portugal. The initial correspondence was on botany and he sent seeds. In a succession of letters in September 1783 he was able to describe the pioneering balloon ascents of the Montgolfier brothers, Charles and others.(18) Later, Montgolfier's assistant, Argand was to travel to London with a letter of introduction from Broussonet, who wrote many more letters to Banks on the subject of ballooning throughout the remainder of 1783 and 1784. (Another correspondent from Paris who provided Banks with detailed accounts of the early balloon ascents in 1783 was Benjamin Franklin.)[19] In March 1785 Broussonet was describing Lavoisier's experiments on the synthesis of water from inflammable air.[20] Later letters discuss agriculture, including the breeding of sheep. By 1790 Broussonet was complaining that French *savants* had largely abandoned science for public affairs.[21] He was himself elected a deputy for Paris in 1791 in the National Assembly but in 1794 he was obliged to flee to Spain. The naturalist the abbé Pourret was another Banks correspondent who was obliged to flee to Spain during the Terror.[22] Before Condorcet became involved in the politics of the Revolution he was the secretary of the Paris Academy of Sciences and in this capacity wrote several letters to Banks. A.L. Jussieu was another correspondent of Banks who explained his failure to correspond in 1792 as due to his political activities.[23]

In the course of the French Revolution many aristocratic refugees fled to England and some of them requested help from Banks. For example, in an undated letter the Marquise de la Bretonnière explained that since coming to England she had been living in a single room and living on the sales of her jewellery.[24] Now that her funds were exhausted she hoped that Banks would help her to travel to Austria where she hoped the Emperor would help her. Slightly more related to science was a letter from a foreign F.R.S., Dominique Cassini in Paris in 1800 explaining that his widowed aunt, who had taken refuge in London, was in distress. Perhaps Banks could obtain help from the King?[25]

Banks was also approached by a number of people whom we now might want to dismiss as cranks. One of these was the obscure French philosopher and author Hyacynthe Azaïs (1766–1845) who had tried in vain in 1806 to interest the First Class of the Institute in his speculations on caloric.[26] Having received little encouragement from the official French body of science, he turned to the Royal Society and wrote to Banks in 1807 about his theory of the earth.[27] In 1809 Azaïs was making further claims to have advanced science[28] but Banks was not able to give him any encouragement either.

A particularly important part of Banks's correspondence relates to the period of the war with France when Banks frequently used his good offices to safeguard the property or persons of men of science. Sir Gavin de Beer had studied much of this story: for example, the account of the loss of the collection of the naturalist La Billardière in 1794.[29] A more notable case was that of the mineralogist Dolomieu, who was captured by the Neopolitans in 1799 and imprisoned. As soon as Banks heard the news he made representations to the British government.[30] He argued that Dolomieu was not to be associated with the French army under Bonaparte but rather considered as a neutral man of science. One of Banks's letters on behalf of Dolomieu was read out at a meeting of the First Class on 10 February 1800. Although Banks failed to obtain Dolomieu's immediate release, the international pressure on his behalf helped improve the conditions under which he was held. Dolomieu was eventually released in March 1801 as a result of one of the conditions laid down by Napoleon for an armistice. The imprisonment of Dolomieu produced a copious correspondence involving Banks and members of the French National Institute and did much to bring together men of science in the two countries despite the war. Banks was showing that he believed that science was above politics.(31) Although a patriotic Englishman, Banks saw himself as one of the leaders of the international scientific community with interests and responsibilities which transcended national quarrels.

An important intermediary in scientific relations with France at this time was Charles Blagden (1748–1820) — Sir Charles from 1792. Blagden's ties with France went back to the *ancien regime*. He had been in Paris in 1792 and was one of those Englishmen who could hardly wait to get back to France when there was the prospect of peace with Napoleon in March 1802. Blagden had been a regular correspondent of Banks since the 1770s. He was a particularly assiduous correspondent from Paris during the whole year-long Peace of Amiens, writing several letters every month to keep Banks informed about scientific news in the French capital.(32) Blagden was given an especially warm welcome by his friends Berthollet and Laplace. He met Bonaparte several times and, being convinced of the genuine interest in science of the First Consul, he suggested to Banks that he should be elected to the Royal Society in recognition of his patronage of science!(33) Needless to say, this suggestion was not taken any further by Banks. Nevertheless, Bonaparte continued to play the role of patron of science and, in June, Blagden reported Bonaparte's plan for a grand prize for galvanism,(34) a prize which was later to be awarded to Humphry Davy.(35) Blagden regularly attended the meetings of the First Class of the Institute and reported on the proceedings to Banks.

The letters written by Banks himself to foreign correspondents, if they survive at all, are naturally dispersed. One of the best surviving collections of letters by Banks to a single correspondent is the collection of Blagden correspondence in the archives of the Royal Society. Although most of the letters by Banks were addressed to Blagden when he was in Britain, many of the later letters from 1814 onwards were sent to Blagden at a Paris address. Several express thanks for the information that Blagden was able to supply about French science.(36)

Returning to the earlier period, immediately after the breakdown of the Peace of Amiens in 1803, although Blagden had been obliged to return to England, Banks was able to continue to correspond with several French men of science including the geologist Faujas de St. Fond, the zoologists Cuvier and Lacépède and the astronomer Delambre. Delambre was also now one of the permanent secretaries of the First Class and in that capacity was able to obtain the release of an English astronomer, John Osbourne F.R.S., who had been interned in France. Banks thanked Delambre and asked for his good offices on behalf of other internees.(37) During the remaining years of the war Delambre was particularly useful to Banks in providing copies of French scientific publications, notably the *Mêmoires* of the First Class and the *Connoissance des Tems*. Delambre planned to use the good offices of the Ministry of the Navy to send such volumes to England by the occasional ship of truce.(38) Banks for his part was able to send the *Philosophical Transactions* and other recent British publications to Delambre for the benefit of members of the First Class.

Although France was the centre of the Revolutionary and Napoleonic wars which cast a shadow over Europe for most of the years 1792–1815, it was not, of course, the only European country with which Banks corresponded. One has to be careful when speaking of the geographical divisions of Europe in this period since in several ways it was significantly different from the national boundaries we may take for granted in the late twentieth century. In particular, Italy and especially Germany consisted of a multiplicity of semi-independent states which did not achieve national unity until the 1860s or 1870s. Also as the French armies advanced into northern Italy, the Rhineland and the Netherlands, a number of new vassal states were set up. Napoleon's brother Louis became King of the Netherlands in 1806, Joseph became King of Naples and Jerome became King of Westphalia, some of these states having previously been organised by the French as republics.

Marc-Auguste Pictet, editor of the *Bibliothèque britannique* and a citizen of the "Helvetian republic" (Switzerland) was one of Banks's correspondents. In fact Pictet spent so much time in Paris that he might almost be regarded as half French. Banks was able to send geological specimens to Pictet's fellow-Genevan, H.B. de Saussure (1740–1799),[39] whose book, *Voyages dans les Alpes* did much to open up this region. From Brussels (not yet part of an independent Belgium) the natural philosopher, abbé Mann (1735–1809) kept up a regular correspondence from 1788 to the end of the century, keeping Banks in touch with the affairs of the Brussels Academy and supplying much miscellaneous scientific news. From Haarlem the electrician van Marum (1750–1837) wrote several letters. His fellow-countryman Boddaert (1730–1796) wrote mainly about botanical and zoological matters.

In the German states the anatomist and early anthropologist Blumenbach (1752–1841) was Banks's most assiduous correspondent. The correspondence, mainly from Göttingen, was principally concerned with expeditions and the natives of different races but there was also considerable exchange of publications and mineral specimens. Another German, Lorenz Crell (1744–1816) from Helmstädt was a member of many foreign scientific societies and began his correspondence with Banks in 1785 by asking to be elected F.R.S.[40] His continuing correspondence, flattery and the sending of successive volumes of the chemical journal of which he was editor eventually had the desired effect and he was elected in 1788 at the same time as Lavoisier and the Danish astronomer Thomas Bugge (1740–1815). Bugge too had solicited membership, helping his case by informing Banks that he had been elected a member of the Royal Society of Science of Copenhagen.[41] Bugge subsequently supplied letters of introduction to Banks for several Danish visitors to England. The Swedish botanist Olof Swartz (1760–1817) was another correspondent, as was the Finnish chemist and mineralogist Gadolin (1760–1852). Also from these northern latitudes was the naturalist Pallas (1741–1811), who wrote from St. Petersburg.

Moving south to Vienna, there is an interesting letter in 1783 from the physiologist Ingenhousz (1730–1799) criticising Priestley, who was his rival to the claim for the discovery of photosynthesis.[42] By 1786 Ingenhousz was explaining to Banks that he had heard about recent dissensions at the Royal Society and for this reason he had preferred to send several of his recent papers to the *Journal de Physique* in Paris.[43] There was also Viennese correspondence from the naturalist N.J. Jacquin (1727–1817) and his son, the botanist, J.F. Jacquin (1766–1839). Both wrote on botany and exchanged seeds and publications. Finally, in the Italian peninsula the most active correspondent was the chemist and naturalist Fabbroni (1752–1822), who lived in Florence[44] and again the dominant subject was botany. The diplomatist John Strange

(1732–1799) spent some time in Venice and wrote from there to Banks over the period 1779–1785. He was able to give Banks detailed information about literary and scientific periodical publications in Italy and also supplied some meteorological observations. Most famous of all, Volta corresponded with Banks from Pavia. He was very grateful to receive the Copley Medal from the Royal Society in 1794 for his memoir on the effect of electricity on muscular movement,(45) the first time it had been awarded to a foreigner. It was no doubt this favourable reception of his work by the Royal Society that prompted Volta to send to the Society his famous paper on the electric pile,(46) which constitutes the beginning of current electricity.

After the Peace of Amiens, which had provided the first opportunity since the outbreak of war for British travellers to visit France, there was only the case of Sir Humphry Davy, who, having won a prize from the First Class of the Institute, was treated as an exception and given a passport by the French in 1813 despite the war. After two months in Paris he travelled to the south of France and Italy.(47) Ordinary mortals, however, had to wait until after the first abdication of Napoleon in April 1814. Sir Charles Blagden was among the first wave of British visitors to France but left when Napoleon's brief return to power was imminent. He was, however, able to make several more visits before his death in 1820.(48) The Scottish physicist Sir John Leslie was another early visitor to Paris. In 1817 there followed Mary Somerville, who met Laplace and later published a popularisation of his *Mécanique céleste*. The visits of John Dalton and Charles Lyell belong to the 1820s. Several important scientists from France visited Berlin in 1817, including Gay-Lussac and Biot; also Humboldt and Arago, who were concerned with geodesic measurements. Arago returned in 1822 to make measurements of terrestrial magnetism. Thus the importance of travel from both sides of the Channel was not only to meet foreign scientists but to carry out important field work.

After more than twenty years of war, there was great curiosity in Britain about all aspects of French life. Those interested in science wanted to learn about the new scientific institutions that had been established in Paris after the revolution. The Royal Academy of Sciences had been succeeded by the First Class of the National Institute, also an elite body in which each of the main scientific disciplines was represented by six members. The former Jardin du Roi had exceptionally not been closed down but reorganised and expanded as the Muséum d'Histoire Naturelle with enlarged collections and an important teaching function. The highest level of physical science was represented in the teaching of the newly-founded Ecole Polytechnique, which, like the Muséum, had leading scientists as its professors. Some of these institutions provided precedents for British science, for example, in the establishment of zoology in the 1820s and in attempts at reform of the Royal Society in the 1830s.(49) Thus influential ideas about the organisation of science as well as scientific data depended on the all-important Channel crossing.(50)

We cannot fail to be impressed by the diversity of Banks's European correspondence. Although known first as a naturalist, Banks had very wide interests which were reinforced by his election as President of the Royal Society. He became the key figure in the politics of English science. His support was vital in securing election to the Royal Society. His social, political and scientific contacts became so extensive that anyone who wanted action on the scientific front was advised to write to Banks. Through his correspondence Banks acquired a detailed and unrivalled knowledge of scientific developments throughout the continent of Europe even in time of war. Although increasingly autocratic in his later years, Banks used well the power that he had acquired in rescuing captives and in proclaiming that the cause of science transcended national quarrels.

Notes

(1) G.S. Rousseau and R. Porter (eds.), *The ferment of knowledge. Studies in the historiography of eighteenth-century science*, Cambridge, 1980.

(2) J.R. Partington, *History of chemistry*, London, 1962, vol. 2, 437.

(3) A. Young, *Travels in France*, 1792; G.E. Mingay, *Arthur Young and his times*, London, 1975, 177–221. For general accounts of British travellers to France see J. Lough, *France on the eve of revolution. British travellers' observations, 1763–1788*, London, 1987 and C. Maxwell, *The English traveller in France, 1698–1815*, London, 1935.

(4) On industrial espionage in Britain see the work of the economic historian, J.R. Harris.

(5) Often eighteenth-century science owed more to technology than technology to science.

(6) T. Kemp, *Economic forces in French history*, London, 1971, chap. 5; C. Trebilcock, *The industrialisation of the continental powers, 1780–1914*, London, 1981, 129–132.

(7) D.M. Bickerton, *Marc-Auguste and Charles Pictet, the Bibliothèque britannique (1796–1815) and the dissemination of British literature and science on the continent*, Geneva, 1986.

(8) W.R. Dawson (ed.), *The Banks' letters. A calendar of the correspondence of Sir Joseph Banks preserved in the British Museum, the British Museum (Natural History) and other collections in Great Britain*, London, 1958, 517–8. For general information on Banks's foreign correspondents see also H.B. Carter, *Sir Joseph Banks (1743–1820). A guide to biographical and bibliographical sources*, London, 1987, 120–6.

(9) There is a total of nineteen of Banks's letters in the archives of the Academy of Sciences. Several of these were to A.L. de Jussieu, Delambre et al.

(10) M. Crosland, *Science under control. The French Academy of Sciences, 1795–1914*, Cambridge, 1992.

(11) See Banks's letter of thanks of 21 January 1802, Archives, Academy of Sciences, Paris, Banks' dossier.

(12) M. Crosland, 'Explicit qualifications as a criterion for membership of the Royal Society: a historical review', *Notes and Records of the Royal Society of London* (1983), **37**, 167–87.

(13) Royal Society, *The Record of the Royal Society of London*, 4th edn., London, 1940, 95.

(14) Natural History Museum, London, Dawson Turner Collection of copies of correspondence of Sir Joseph Banks, vol. 6, 40.

(15) He was only elected in 1829 when he was over 80 years old.

(16) British Museum [hereafter B.M.] Add. MS. 8096, 465.

(17) Carter, op. cit. (8), plate 11. No distinction is made, however, between foreign and British correspondents in this enumeration.

(18) B.M. Add. MS. 8095, 134–9.

(19) Royal Society, Letters and Papers, Decade 8, 36, 37.

(20) B.M. Add. MS. 8096, 29–31.

(21) B.M. Add. MS. 8097, 273–4, 281–2.

(22) B.M. Add. MS. 8098, 374–5.

(23) B.M. Add. MS. 8098, 149.

(24) B.M. Add. MS. 8098, 230.

(25) B.M. Add. MS. 8099, 32; cf letters from Barruel to Banks, B.M. Add. MS. 8099, 8–10.

(26) Crosland, op. cit. (10), 340.

(27) Banks's correspondence in possession of Lord Brabourne, 81–4.

(28) B.M. Add. MS. 8100, 38, 41.

(29) Sir Gavin de Beer, *The Sciences were never at war*, London, 1960, chap. 5.

(30) Ibid., chap. 7.

(31) See ibid., 197 for Jenner's remark that "the sciences were never at war". The same comment was later made by Humphry Davy, J.A. Paris, *Life of Davy*, London, 1831, vol. 1, 261.

(32) Dawson, *op. cit.* (8), 86–93.

(33) Letter of 25 May 1802, B.M. Add. MS. 33272, 187–8.

(34) B.M. Add. MS. 33272, 195–6.

(35) M. Crosland, *The Society of Arcueil. A view of French science at the time of Napoleon I*, London, 1967, 20–5.

(36) E.g. Banks to Blagden, 16 October 1814, Royal Society, Blagden letters, BLA.b.51.

(37) Letter of 30 January 1804, Natural History Museum, London, Dawson Turner Collection, vol. 14, 196–8.

(38) B.M. Add. MS. 8100, 27–8.

(39) B.M. Add. MS. 8098, 84–5.

(40) B.M. Add. MS. 8096, 64–5.

(41) B.M. Add. MS. 8096, 13–14.

(42) B.M. Add. MS. 8096, 114–5.

(43) B.M. Add. MS. 8097, 83–4.

(44) Dawson, op. cit. (8), 310–5.

(45) M.Y. Bektas and M. Crosland, 'The Copley medal: the establishment of a reward system in the Royal Society, 1731–1839', *Notes and Records of the Royal Society of London* (1992), **46**, 43–76.

(46) A. Volta, 'On the electricity excited by mere contact of conducting substances of different kinds' (in French), *Philosophical Transactions* (1800), **90**, 403–31.

(47) An account of Davy's travels is given in B. Bowers and L. Syms (eds.), *Curiosity perfectly satisfyed. Faraday's travels in Europe, 1813–1815*, London, 1991.

(48) Crosland, op. cit. (35), 417.

(49) For French science as a model see C. Babbage, *Reflections on the decline of science in England*, London, 1830, e.g. 32–6.

(50) For the importance of the Paris hospitals for the advancement of British medicine see R.C. Maulitz, 'Channel crossing: the lure of French pathology for English medical students', *Bulletin of the History of Medicine* (1981), **55**, 475–96.

B. Jonsell, 'The Swedish connection', in *Sir Joseph Banks: a global perspective* (eds. R.E.R. Banks and others), Royal Botanic Gardens, Kew, 1994, 23–29.

3. THE SWEDISH CONNECTION

BENGT JONSELL

Bergius Foundation, Royal Swedish Academy of Sciences, Stockholm, Sweden

The very first occasion which made Joseph Banks's name well known in Sweden, or rather to one very highly influencial person in that country, was to the best of my knowledge a letter from John Ellis to Linnaeus in August 1768. It is true that the Swede Daniel Solander, who by then had been in Britain for a number of years, knew Banks very well, but being a poor correspondant with his old teacher in Uppsala, Banks's name does not seem to have reached Linnaeus that way. However, John Ellis, for long an intimate correspondant and great admirer of Linnaeus, revealed the following information.

"I must inform you that Joseph Banks Esquire, a gentleman of 6000 per annum estate, has prevailed on your pupil Dr Solander to accompany him in the ship that carries the English astronomers to the newly discovered country in the South Sea. They are to proceed ... on further discoveries of the great Southern continent ... No people ever went to sea better fitted out for the purpose of Natural History, nor more elegantly ... In short, Solander assured me this expedition would cost Mr Banks 10000 pounds". Ellis had then taken leave of Banks and Solander three days earlier. As an act of devotion to Linnaeus Ellis added: "All this thing is owing to you and your writings."

Whether really true for the South Sea expedition those words of Ellis are doubtless true as to Banks's connections with Sweden. Directly or indirectly Linnaeus is behind them all, from the friendship with Solander, over the affairs concerning the Linnaean collections, the visits of the long array of Swedish naturalists to Soho Square and the relations with the Swedish Academy of Sciences all the way to his memories of Jonas Dryander in 1810, ten years before his own death. I shall here give some glimpses from all these fields.

Linnaeus's influence is everywhere felt and the *ultima causa* is to be seen in the month he spent in England in 1736, when he at last made such a deep impression on sceptical British scientists such as Sir Hans Sloane (1660–1753) and Johann Jacob Dillenius (1684–1747). Linnaeus became acquainted also with Peter Collinson (1694–1768), who jointly with his merchant colleague John Ellis (1711–1776) were to become Linnaeus's strongest supporters in Britain, actively urging Linnaeus to send one of his disciples to spread his method of classification further. This disciple was Daniel Solander (1733–1782), who had arrived in 1763 and immediately started to provide useful service to the British Museum where in the autumn of 1764 he came to know Banks. It is highly likely that Banks was both taught and inspired by Solander. Thus Collinson's and Ellis's hopes bore fruit quickly — to the extent that travel to Sweden to see Linnaeus and learn from him in person was discussed. From here is the expression about Banks as "the best disciple Linnaeus never had". Nevertheless, through talks about the Endeavour expedition Banks in turn came to inspire Solander, who finally could not help asking, "Would you like a fellow traveller?" — "Someone like you would give untold treasures and rewards" was Banks's reply, and from that moment we may see Solander as a naturalized Englishman.[1,2] What they all had in common is from then on hardly a connection with Sweden.

The South Sea expedition, however, gave Linnaeus little rest, and for the few years of his life he had left the thoughts of those marvellous plants which he never came to see and never had the opportunity to describe himself, troubled his connections with the Banksian circle.

Linnaeus had good reasons to expect information about the results of the expedition. There is a long, interesting letter from Solander to Linnaeus sent from Rio de Janeiro on the 1st December 1768 on the outward leg of the expedition. This letter is not amongst the Linnaean collection in London but was retained in Sweden, probably because it also deals with intimate family matters, and was discovered only in the 1930s.[3,4] In this letter Solander promised after the return from the voyage to pay a visit to Uppsala together with Banks. Far from that, however, not even a single specimen reached Linnaeus, whose disappointment was openly shown in a letter to Ellis. "If he had brought some specimens with him I could at once have told him what were new; and we might have pored over books together, and he might have been informed or satisfied upon subjects, which after my death will not be so easily explained". It had been Ellis, again, that first reported to Linnaeus on the successful return of the expedition "laden with the greatest treasures of natural history ever brought into any country at one time by two people", to which Linnaeus had replied that he had never received a more welcome letter, and "if not bound here by 64 years of age and a worn out body, I would this very day set out for London to see this great hero of botany".

Besides Ellis, two Linnaean disciples temporarily in London, Anders Berlin (1747–1773) and Henrik Gahn (1747–1816) informed Linnaeus about the South Sea treasures, and after 10 months Banks signed a letter, apparently written by Solander, very polite and with many excuses and explanations why the specimens reserved for Linnaeus had not been despatched. After Anders Berlin had revealed to Linnaeus that a new expedition was planned with Banks and Solander taking part, the despairing lines cited above were written.[5]

The affair did not end with Linnaeus's death in 1778. That year Linnaeus filius had asked for permission to include the South Sea plants in his Supplementum Plantarum, which was to appear in 1781, a request politely but firmly declined.[6] During the 1770s a certain amount of tension had developed between the two Linnaeuses and the Banksian circle. It is not easy to find out how serious it really was, but it might have a background in the disparaging view of Linnaeus filius held by many Swedish colleagues of the elder Linnaeus in his days of vigour. A leading adversary of Linnaeus filius was the former disciple of the elder Linnaeus, the physisician and Lapland traveller Lars Montin (1723–1785), who corresponded frequently with Banks in the 1770s. He was moreover the uncle of Jonas Dryander (1748–1810). The two relatives kept close contacts even after Dryander's arrival in London in 1777 and certain criticisms against the younger Linnaeus were apparently spread around.

The unfortunate episode concerning the name *Banksia* may have contributed to this situation. The genus that was originally given that name by the elder Linnaeus in his manuscript to Supplementum Plantarum was included within a previously erected genus by the younger Linnaeus. This was resentfully received in the Banksian circle — though no reaction from Banks personally is known. The matter was soon resolved, however, when Linnaeus filius used the name *Banksia* for another genus discovered by the Forsters during the second Cook expedition — the beautiful member of Proteaceae with which we are familiar to today.

Nevertheless, even when Linnaeus filius, in his capacity as professor in Uppsala, visited London during the 18 months from April 1781 to August 1782 the animosity over the South Sea plants was still apparent. Thus Dryander wrote to Thunberg in Uppsala on the 3rd of July 1781 that "Nobody will yet receive any of Sir Joseph's South Sea plants besides myself, and least of all Linnaeus..", and about half a year later: "Professor Linnaeus will return to Sweden without having touched Mrss Banks's and Solander's South Sea plants; he has not even seen them. We have not worked upon them since he arrived here and will not as long as he stays".(7) This is much in contrast to how Carl Peter Thunberg (1743–1828) had been received a few years earlier when he arrived at Soho Square on his way home to Sweden from the Cape and Japan. He was given full access to the Endeavour collections and he praised the place as an Academy of Natural History. Apparently he rewarded the hospitality with material and information from Japan and may have inspired Banks to undertake the publication of the Icones Kaempferianae — Kaempfer being Thunberg's only botanical predecessor in Japan.(8)

It might here be added in passing that some fragments of the South Sea collections reached Sweden, but they were from among the conchyles housed by Banks, Solander and the Duchess of Portland. Johan Alströmer, a brother of the elder Linnaeus's two close friends Jonas and Claes, was in London in 1777–78 and in close touch with both Banks and Solander. He was permitted to take with him samples from the collections mentioned but they were most probably already destroyed in the fire of his home town Alingsås in 1779.(9,10)

In spite of a few embarassing episodes, and I will come back to one more shortly — that concerning the Linnaean collections — Linnaeus filius seems to have had a good time in London, the best year of his life as Arvid Uggla thought. His position in Sweden was, as I indicated, highly troublesome. Of all the Swedes received at Soho Square he was at his visit the one at highest rank, a full university professor. He was recommended to Banks by a letter from the above-mentioned Johan Alströmer. A letter to his close friend Claes Alströmer a fortnight after his arrival discloses his nervousness about how he might be received by Banks but also the nice surprise it was to have been so extremely well treated right from the beginning. However, it may by no means have been easy for him to appear as the son of Linnaeus — his inevitable fate — and even to take part in the opulent life of society. He was certainly not a Solander about whom Boswell wrote "Throw him wherever you wish, he swims". He had never been abroad before and now saw his travel money shrink precariously because of the life of an unexpectedly high standard in which he had to take part. In spite of that and the difficulties touched on above, he seems to have managed well in the long run, and learnt a lot. A blow was, however, Solander's death that occurred during his stay, in May 1782. Regrettably he had himself only little more than a year left to live after his return to Sweden.

Linnaeus filius's untimely death brought to the fore the revival of a business that had been already discussed five years earlier — the fate of the Linnaean collections. Banks had here played an active role. After the elder Linnaeus had died in 1778 Banks approached Lars Montin, as Dryander's close relative and an ardent plant collector, Banks's most natural contact in Sweden at the time. Banks offered via Montin a sum of 1000 guineas for the Linnaean herbarium, which Linnaeus filius called "a cruel offer".(11,12) Thanks to the circumstance that the heirs, as well as the younger Linnaeus, and his mother and sisters, did not know about Banks's offer before a treaty was reached among them, the collections this time stayed in Sweden as the property of Linnaeus filius. After his death in 1783 the matter came up again, now ending with the result which we know.(13)

The heirs were now decided to make as much money as possible out of the collections and engaged a friend of the family Dr Johan Gustaf Acrel (1741–1801) to handle the transaction. Of course Banks's interest was recalled and not surprisingly Montin was asked to act in this business via his nephew Dryander. He did, but meanwhile Acrel approached a colleague of his, Dr Johan Henrik Engelhardt (1759–1832), now for some time in London, asking him also to approach Banks, and it was Engelhardt's letter that arrived on that famous breakfast of 23rd December 1783 when Banks handed over the offer to James Edward Smith, who happened to be present. Two special circumstances seem to have played a role in the course of events. Economic strains and other difficulties at that time(14) made Banks not personally interested in buying the collections, at least not on that very occasion. James Edward Smith knew Engelhardt from their common studies in Edinburgh. Banks played his important role as mediator of the transaction, a conception leading later to what was to become the Linnean Society of London, but the continuation is another story.

One thing should, however, be added. Actions were taken in Sweden to prevent the collections being sold abroad. Thunberg, now professor and successor to the two Linnaeuses, as soon as he learnt about Acrel's step also acted along the line Montin–Dryander–Banks to persuade the latter not to buy at all or at least to refrain from purchasing the plants Linnaeus the younger had obtained from Banks among others during his stay in London. It seems possible that Thunberg's action might have made Banks declare that he did not want to buy the Linnaean collections if there were purchasers in Sweden, because he thought they should not leave that country. If not, he would like to be as close at them as any other foreigner. So Banks's opinion is quoted in a letter from Montin to Thunberg of 26th December 1783, making it clear that Banks accordingly knew what was afoot well before the breakfast of the 23rd. Thus, the offer arriving via Engelhardt was not a total surprise.

I have mentioned a number of Swedes, Linnaean disciples, who benefited from Sir Joseph in London. Two more must find place here, Adam Afzelius (1750–1837), and Olof Swartz (1760–1818). Afzelius was a pupil of the aged Linnaeus and the last surviving of them all when he died in 1837, almost 60 years after his teacher. His inclination to Britain was largely because he was an ardent Swedenborgian but he enjoyed the friendship and esteem of Sir Joseph, when he first appeared in 1789. He translated Thunberg's travels into English and stayed for three years before he set out for Sierra Leone with a fellow Swedenborgian, but obviously affiliated to the programme of the African Association partly inspired by Banks. In spite of weak health and various hardships including a French attack on Freetown he endured but for a short break until 1796 whereafter he spent another three years in London. Then he returned to what he had called "The gloomy Uppsala"; even when down in Africa he had dreamt of being back there.[15]

Olof Swartz (1760–1818) was just too young to have studied with Linnaeus himself, but stands in the Linnaean tradition. It was on his way back from the West Indies, in particular Jamaica, that he arrived in London in the autumn of 1786 and stayed until the following summer. Swartz is constantly described as the nicest of persons, and he made a deep friendship with Banks as well as with Sir William Chambers and James Edward Smith — the latter called him "the best botanist I have seen since Solander". His work with Banks, his library and his collections was of the utmost importance for the completion of his great works on West Indian plants.[16]

Sir Joseph's affection for Swartz as well as for Solander was unchanged by time, a fact which is proved by a very special circumstance. In 1792 Banks learnt that a relative of the late Solander, a widow of a vicar in Solander's birth town Piteå far in the north was in a precarious economic situation. He therefore sent £250 sterling to the present vicar of the parish to be used so that the widow, to whom the donor's name should be kept secret, obtained the interest thereof for the rest of her life time. And so he states: "After the death of Mrs Idman I wish that the capital is given to the Royal Swedish Academy of Sciences who should administer it to increase the salary for professor Bergianus. My friend Olof Swartz, who is now professor Bergianus is young and Mrs Idman at age, so according to the general course of human life I may hope that he at growing age will enjoy some addition to his salary. The interest of the capital should forever be used as an addition to the salary of the professor on the chair which was installed by Dr Bergius in Stockholm." This has been fulfilled, and today it is the person writing this chapter who has to express the everlasting gratitude to Sir Joseph for the so-called Banksian legacy.

We have already met the Swede who was to serve, effectively and silently, the longest time of all on the Banksian staff, Jonas Dryander. He arrived with a grant from Sweden in 1777 and after Solander's death he became librarian and curator, and he, too, became a naturalized Englishman the details of whose efforts can no longer be considered as a link with Sweden. But he was also a pupil of Linnaeus, and in the Linnaean tradition he was behind most of the descriptions and formal taxonomy in what is generally cited as Aiton's Hortus Kewensis. His great enterprise, the five volume "Catalogus bibliothecae historico-naturalis Josephi Banks..", completed from 1796–1800, must also be mentioned and in this he was for some time assisted by the Swedes Samuel Törner and Fredrik Schutzen. The 19th century bibliographer Pritzel regarded it as the masterpiece among the library catalogues of its century.

There is a letter from Banks to an unknown adressee written in the autumn of 1810, a little more than a week after Dryander's death in which Banks's esteem and friendship are obvious: "Alas, my dear Sir, I have lost my Poor Friend Dryander and in him have lost my right hand, he was the Fountain from whence a stream of my best and most satisfactory amusement flowd in an uninterrupted course across which a Perpetual Dam is now I fear thrown, my hairs are gray and I am not far distant from the haven of Rest, which I look to with hope rather than apprehension".[17] Du Rietz[18] cites further letters with proofs of Banks's affection for Dryander.

So Banks's last personal link with Sweden was gone. We have seen that the connections to that country were multifarious. I have not touched among those which may be self evident and shared with others — his membership of the Swedish Academy of Sciences. The British members were, however, remarkably few among the foreigners, in those years only about a dozen or 10%, much below the number of Frenchmen and Germans.[19] Banks was elected in 1773, and later to three other Swedish learned societies, in Gothenburg, Uppsala, and Lund. He was probably among the more active of the foreign members, not only sending the legacy but a number of books throughout the years, and obtained of course quite a lot in return. The last letter was written from our Academy to Banks by the famous chemist Berzelius, a man of a new era, secretary after Swartz, on the 15th of June 1820, which was 4 days before Banks's death at Spring Grove.

Banks and Sweden is a rich subject, of importance for the development of science. It ought, as Rolf Du Rietz[20] remarked more than 20 years ago, to become a Swedish contribution to research on Banks but it is still waiting for its student. That which is presented here is not much more than the headlines.

Notes

(1) B. Jonsell, 'Linnaeus and his two circumnavigating apostles', *Proceedings of the Linnean Society of New South Wales* (1982), **106**, 1–19.

(2) R. Granit, 'Banks och Solander — två vänner i 1700-talets London', in *Utur stubbotan rot* (ed. R. Granit), Stockholm, 1978, 41–59.

(3) A. Grape, 'Linnaeana bland J.G. Acrels papper i Karolinska Institutets bibliotek', *Svenska Linné-Sällskapets Årsskrift* (1933), **16**, 112–20.

(4) A.H. Uggla, 'Daniel Solander och Linné', *Svenska Linné-Sällskapets Årsskrift* (1955), **37/38**, 23–64.

(5) Ibid.

(6) Ibid.

(7) A.H. Uggla, 'Från Linné den yngres Englandsresa', *Svenska Linné-Sällskapets Årsskrift* (1953), **36**, 71–7.

(8) H.B. Carter, *Sir Joseph Banks*, London, 1988.

(9) R. Du Rietz, 'Sir Joseph Banks — en litteraturöversikt', *Lychnos* (1962), 200–11.

(10) S. Rydén, 'Johan Alströmer och Daniel Solander — två Linnélärjungar i London', *Svenska Linné-Sällskapets Årsskrift* (1960), **43**, 53–71.

(11) A.H. Uggla, 'Linné den yngres brev till Abraham Bäck 1778', *Svenska Linné-Sällskapets Årsskrift* (1957), **39/40**, 138–65.

(12) A.H. Uggla, 'Linné den yngres brev till Abraham Bäck. 2. 1779–1783', *Svenska Linné-Sällskapets Årsskrift* (1958), **41**, 61–100.

(13) T.M. Fries, *Linné. Lefnadsteckning*, Stockholm, 1903, vols. 1, 2.

(14) Carter, op. cit. (8).

(15) C.G. Widstrand, 'A note on Adam Afzelius', in 'Adam Afzelius Sierra Leone Journal 1795–1796' (ed. A.P. Kup), *Studia Ethnographica Upsaliensia* (1967), **27**, xi–xv.

(16) B. Nordenstam, 'Olof Swartz' in *Bergianska botanister* (ed. B. Jonsell), Stockholm, 1991, 23–43.

(17) N. Svedelius, 'Ett vackert vitsord om Jonas Dryander av Sir Joseph Banks', *Svenska Linné-Sällskapets Årsskrift* (1946), **29**, 109–10.

(18) R. Du Rietz, 'More light on Jonas Dryander', *Svenska Linné-Sällskapets Årsskrift* (1964), **47**, 82–3.

(19) S. Lindroth, *Kungl. Svenska Vetenskapsakademiens historia 1739–1818*, Stockholm, 1967, vols. 1, 2.

(20) Du Rietz, op. cit. (9).

A. Agnarsdottir, 'Sir Joseph Banks and the exploration of Iceland', in *Sir Joseph Banks: a global perspective* (eds. R.E.R. Banks and others), Royal Botanic Gardens, Kew, 1994, 31–48.

4. SIR JOSEPH BANKS AND THE EXPLORATION OF ICELAND

ANNA AGNARSDÓTTIR

University of Iceland

In 1772 Sir Joseph Banks set off to explore Iceland, leading the first scientific expedition undertaken by foreign naturalists to that country. At the time Iceland was a dependency of Denmark. As a consequence of that visit, Banks became the acknowledged British expert on Iceland and a faithful friend of the Icelanders. Three decades later, during the Napoleonic Wars, Banks assumed a crucial political role as self-appointed proctector of Iceland.

The Iceland Expedition[1]

It was in the year 1772 that Banks's connection with Iceland began. Only 29 years of age, this naturalist explorer had already been to Labrador and Newfoundland as well as sailing to the South Seas on the *Endeavour* with Captain Cook in 1768–71. Due to the success of this voyage another expedition to the South Pacific was planned for the following year, and Banks was again invited to join. As before, he assembled a party of scientists, draughtsmen and assistants at his own expense. Among them were the Swedish botanist Daniel Solander, who had accompanied him on the *Endeavour* voyage, and the Scottish physician James Lind, who was interested in astronomy. The artists were the Miller brothers —John Frederick and James — and John Cleveley jr. However, Banks was displeased with the shipboard facilities and after a dispute with the Navy Board withdrew from the voyage. Banks was "disagreeably disappointed",[2] but, wishing to employ his men "in some way or other to the advancement of Science",[3] he "resolved upon another excursion".[4] By early June Banks had decided to head north, his choice falling on Iceland.[5] Thus instead of the South Pacific, Iceland became his destination. The question is why?

Banks wrote in his journal that as the sailing season was much advanced he

"saw no place at all within the Compass of my time so likely to furnish me with an opportunity as Iceland, a countrey which ... has been visited but seldom & never at all by any good naturalist to my Knowledge the whole face of the countrey new to the Botanist & Zoologist as well as the many Volcanoes with which it is said to abound made it very desirable to Explore it ...".[6]

Accordingly, Iceland had the advantages of being relatively near and unexplored, its chief attraction being that it was full of volcanoes. In his passport, quickly issued on 2 July by Baron von Diede, the Danish envoy in London, the main purpose of Banks's visit was recorded as "observing Mount Hekla",[7] the most famous of the Icelandic volcanoes. Two Icelandic scholars have suggested that Solander may also have influenced his decision. The Swede was almost certainly acquainted with the Danish botanist Johan Gerhard König, both being pupils of Linnaeus, and been familiar with the fact that König had been sent to Iceland to collect plants in 1764–65 for the *Flora Danica.* Furthermore, a couple of months earlier, in April 1772, Banks had been given some specimens of Icelandic lignite (*surtarbrandur*). This would have strengthened his belief that Iceland had much to interest a naturalist.(8)

Banks prepared his voyage as best he could within the limited period of time he had. Apparently he found no-one in London who had been to Iceland, but Clause Heide, a Dane resident in London, gave Banks information "chiefly out of books". Heide promised to write to Denmark to find the best advice on "the most proper places for you [Banks] to go to view Mount Heckla, or what place is burning at present, to give the names of people of the greatest note, likewise letters of recommendation ...".(9)

Yet another Swede joined Banks' party — Uno von Troil, a friend of Solander's, who being interested in the Icelandic language, was invited to join the expedition at the last moment. The Swedes would serve as interpreters. Von Troil had come to England directly from Paris, where he had met such luminaries of the Enlightenment as Rousseau, d'Alembert and Diderot.(10) On the eve of departure the Banks Expedition bound for Iceland was about twenty strong.

Banks's chartered ship, the *Sir Lawrence* eventually left Gravesend on 12 July, ironically the same day as Cook started on his second voyage. They sailed to Iceland by way of the Western Isles, making many leisurely stops on the way, and, after suffering extreme bouts of seasickness, finally arrived at their destination — south-west Iceland — on 28 August. — This was very late in the season.

In the late 18th century Iceland was ruled by Denmark, with the King's representative, a governor (*stiftamtmadur*) resident in the island. Iceland had a population of about 50,000. Society was made up of a small landowning class and a large tenant peasantry. Most Icelanders lived on isolated farms and were primarily engaged in sheepfarming, with fishing as a subsidiary occupation. There were no villages, only trading stations dotted around the coast. The Iceland trade was conducted as a monopoly with the Danish merchants sailing to Iceland in the spring bringing necessities to an island with few resources, trading during the summer, returning to Denmark in the autumn. The Icelanders were strictly forbidden to trade with foreigners at any time. The level of education, as Banks was to note, was higher in Iceland than in most other European countries at the time,(11) literacy being widespread, and the sons of the élite were educated at the University of Copenhagen.

The arrival of the *Sir Lawrence* initially caused great alarm. Many boats were out fishing. But as the brig approached, the Icelanders refused to come near them, probably fearing a repetition of a bloodthirsty raid made by Barbary corsairs in 1627. Finally a boat was lowered and a chase began, with the Icelanders rowing furiously away. However, the British managed to overtake them and three Icelanders, coaxed by the Swede Solander, were induced to come aboard the *Sir Lawrence*, visibly trembling even after a large glass of brandy each. Having eaten and drunk "plentifully" and having received Solander's

FIG. 1. An illustration by John Cleveley jr. “View of the Danish storehouses where we lived at Hafnarfiord” is its caption. Hafnarfiord is now correctly spelled *Hafnarfjördur*.[69] To the left Dr. James Lind can be seen making scientific observations[70] and in the foreground are some Icelanders and two rowing boats. The boat to the right is from the *Sir Lawrence*, which chased the Icelanders’ rowing boat as described in the text. They are surrounded by lava.
Reproduced courtesy of the British Library.

guarantee that they were indeed Christians, they recovered — one of them agreeing to pilot the ship into the port of Hafnarfjördur,[12] where Banks found to his "great Joy ... that the sides of the Harbour were constituted of the surface of an ancient flow of Lava".[13]

The governor of Iceland, Lauritz Andreas Thodal, received them with great politeness and friendliness, the Icelanders being "heartily glad" to find they were "Peaceable people". On hearing they were British, the Icelanders had been

"much alarmd at first they thought that we were come with a hostile intention ... they thought that we were the Prelude of an English fleet sent to take possession of the Island ...".

as Banks noted in his journal. Diplomatic relations between the Danish and British courts were strained at the time (due to the marital difficulties of Christian VII and his wife Caroline Matilda, sister of George III). In order to make a good impression on the Icelanders Banks ordered his men to dress in their best apparel and attend church the following Sunday, where as Banks wrote they managed to behave "with all moderation & decency" in spite of the rather unmusical hymn-singing of the Icelanders.

Empty warehouses, belonging to departed Danish merchants, were opened for the benefit of the English visitors, where they settled relatively comfortably for the duration of their visit. The Icelanders came bearing gifts and received in return presents of ribbons or tobacco. Dr. Lind was kept busy dispensing medicine and treated the bewildered Icelanders to electric shocks.

Though they were "sorry to hear that no volcanoes were now burning", the governor assured them they could examine effects of former eruptions everywhere.[14] Banks spent exactly six weeks in all in Iceland, most of the time making short excursions from his base at Hafnarfjördur which, situated in the middle of a lava field, boasted of much to explore. That is where he picked up his famous ballast of lava.[15] One of the excursions was "to see some hot Springs at a place called Reikavik".[16] After exhausting the possibilities of their neighbourhood they eventually, on September 18th, set off on their main excursion to Mount Hekla. Guides were hired and nineteen packhorses loaded with supplies.

First they made their way to Thingvellir, the site of the old Icelandic parliament — the *Althingi* — where they discovered interesting lava formations, continuing on their way to view the warm springs at Laugarvatn. There Banks spent a "Cold & miserable" night, though his landlord was "most Civil", killing a sheep in their honour and heating some milk for supper. Then on to the Great Geysir. Here Banks found his volcano in eruption, but it was a "volcano of water" not of lava. They spent a lot of time measuring the frequency, height and temperature of Geysir.[17] The novelty of boiling meat and fish in the geysers encouraged them to do some cooking, each member of the party having a taste of a ptarmigan, shot by Banks, and almost "boiled to pieces in six minutes" but tasting "excellently" wrote von Troil.[18]

Then on to Skáholt, the bishopric of southern Iceland, where they met the learned bishop Finnur Jónsson, deemed by Banks to be both "venerable" and "sensible", and the master of the cathedral school Bjarni Jónsson who composed odes in Latin in their honour.[19] They also met the "pastor emeritus"

Fig. 2. The inside of a warehouse in Hafnarfjördur where Banks's party stayed while in Iceland. The original of this illustration from the 1772 visit is held by the Linnean Society. Reproduced courtesy of the Linnean Society.

FIG. 3. An illustration by John Cleveley jr. "View of the eruption of Geysir".(71)
Reproduced courtesy of the British Library.

who as Banks noted in his journal was "drunk!", not surprisingly perhaps, as breakfast had consisted of roast mutton and cold cheese served with coffee, wine and brandy.[20] On his departure Banks presented the bishop with a silver shaving knife and Bjarni Jónsson received a silver watch.

And thus finally on to Hekla: it was freezing and the ascent was difficult. Then "a change from cold to warmth ... increased very much and when we arriv'd at the top ... our clothes which were before so much frozen were now thaw'd and became so wet that they were very uncomfortable" as James Roberts, one of Banks's collectors, wrote in his journal. The wind was so strong they were afraid of being blown "into some of the dreadful craters" and so "having finished our observations we thought it safest to descend with all convenient speed".[21] Banks was elated, "no one was ever higher of gentlemen" he wrote in the belief they were the first to reach the summit.[22] Having accomplished the main purpose of their visit the "Philosophical Adventurers" [23] returned to their base in Hafnarfjördur on 28 September, inspecting some more hot springs on the way.

During their stay in Iceland they also did some fishing and botanizing, and led an active social life, dining with the governor and other local notables, the deputy-governor Ólafur Stephensen becoming a particular friend. Banks made a point of entertaining all the leading men of Iceland to exquisite meals prepared by his French Chef Antoine Douez. The Icelandic guests were surprised at the variety of wines and were amazed by the music of the French horns which accompanied these dinners. Afterwards Banks and his men found it "quite astonishing" to see both men and women fearlessly galloping off over the rough beds of lava on their sturdy Icelandic ponies.[24]

Their hospitality was amply returned by the Icelanders. Especially memorable must have been a dinner given by Bjarni Pálsson, Iceland's first state physician, whom they had requested to entertain them "after the Icelandic manner". After starting off with a glass of Danish akvavit (corn-brandy) they had biscuit, cheese, sour butter and dried fish, followed by roast mutton, meat-broth and eventually trout. Bjarni's guests ate with an excellent appetite until the last course was served. Uno von Troil described it thus: "So elegant an entertainment could not be without a desert; and for this purpose some flesh of whale and shark was served. This ... looks very much like rusty bacon, and had so disagreeable a taste that the small quantity we took of it, drove us from the table long before our intention".[25] Bjarni Pálsson gave them gifts of mineral specimens and manuscripts and in return received a magnificent magnifying-glass.[26]

On October 8th they made their farewells and a week later they were in the Orkneys.

Assessment of the Expedition

In the *New Oxford History of England* published a few years ago Banks is portrayed as an arrogant "primadonna" having "initially outshone Cook in fame" then being forced to go off on a "less than successful Icelandic voyage".[27] Was the Iceland voyage a scientific fiasco?

Not being a naturalist I cannot comment on its scientific merits but it seems clear that the Iceland expedition does not compare with the *Endeavour* voyage.

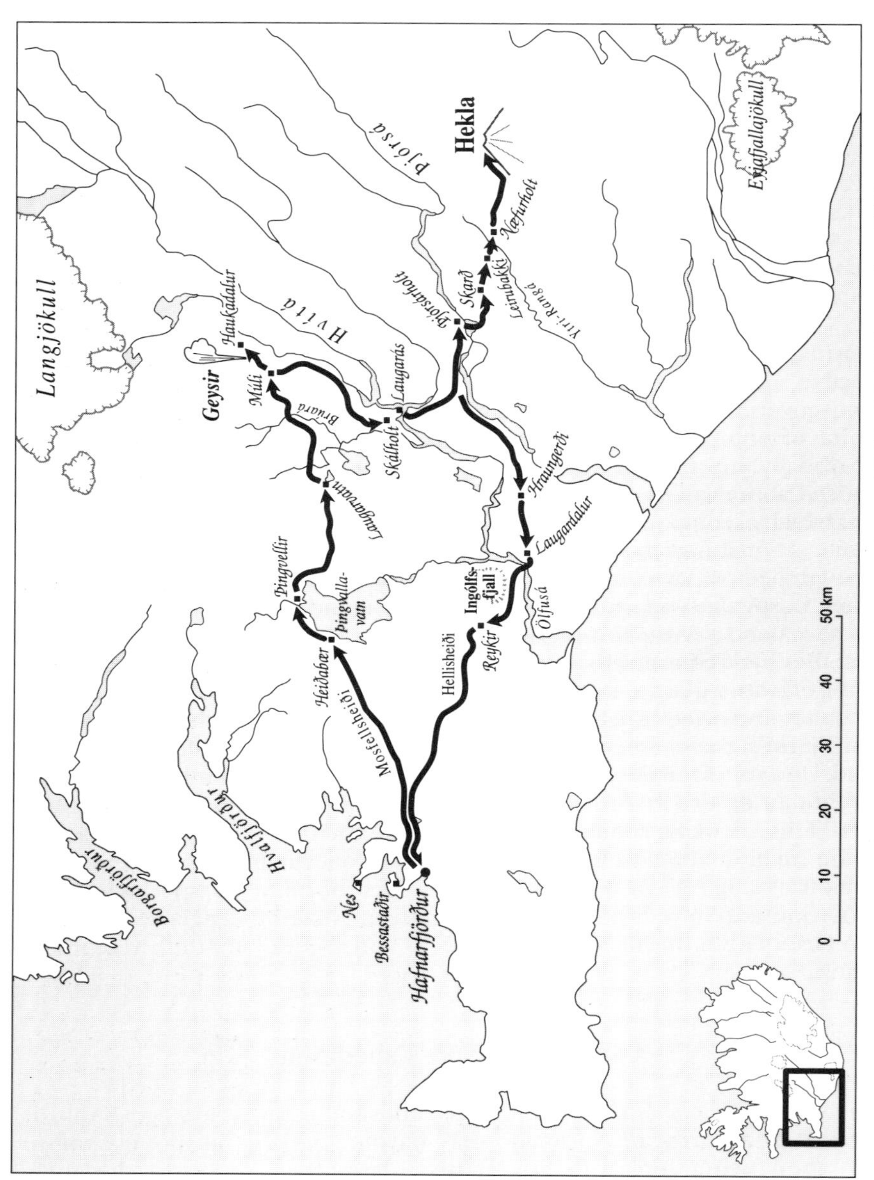

FIG. 4. Map showing the route taken by Banks on his excursion to Hekla, 18–28 September 1772.

From Banks's point of view the tour must in some ways have been disappointing. No volcanoes were in eruption — Hekla had last erupted in 1766 and on average only did so twice a century, so Banks had just missed one! These keen botanists found "but few plants from the lateness of the season",[28] and Bjarni Pálsson — who served the shark and whale — must have told them that he himself had been the first to climb Hekla some twenty years before. And since then he had even been up a second time. In fact, he had been sent with another Icelander, the naturalist Eggert Ólafsson, all over Iceland during the years 1752–7 on an official expedition by "Order of His Danish Majesty", sponsored by the Danish equivalent of the Royal Society to explore Iceland's nature in all its aspects (flora, fauna, geology, meteorological measurements etc.).[29] They too were familiar with Linnaeus, and Pálsson had published works on botany. Both had degrees from the University of Copenhagen. Their description of Iceland was published in two massive volumes, over 1100 pages long, the first volume appearing that very year 1772 in Danish, to be translated into German in 1774 and French in 1802. An abridged version of their *Travels in Iceland* was finally published in English in 1805, where they are known as Olafsen and Povelsen. This was far superior to what Banks could ever have hoped to achieve after only a six-week stay in only the south-west of Iceland. In all fairness to Banks he was sensible to this. He wrote in his journal that despite the late season he had hoped "that something might be done at least hints might be gathered which might promote examination of it by some others".[30] In this he was successful, as other Englishmen were to follow in his footsteps to Iceland: Sir John Thomas Stanley in 1789, William Jackson Hooker the botanist in 1809 and Sir George Steuart Mackenzie, the Scottish mineralogist, in 1810. They all sought his advice and assistance, acknowledging their debt to his pioneering effort.

Though Banks himself published nothing, his companion Uno von Troil, later archbishop of Uppsala, published his book *Letters on Iceland* in Swedish in 1777. It was based on the expedition but von Troil had also consulted other books on Iceland, including the Ólafsson-Pálsson volumes. The *Letters on Iceland* was promptly translated into German (Leipzig, 1779, Nurnberg 1789). No less than three editions were published in English in 1780 (two in London and one in Dublin) with subsequent translations into French in 1781 and Dutch in 1784. Thus it received a wide circulation in Europe.[31] Mackenzie found the book had awakened "the curiosity of science to that neglected, but remarkable country".[32] In the opinion of one modern expert, Von Troil's book is one of the best accounts written on Iceland[33] and, along with Banks's and Roberts's journals, are important sources for Icelandic history today. The illustrations by Cleveley and the Millers, over 70 of them, deposited in the British Museum are invaluable depictions of Icelandic life and in steady use.[34]

Banks moreover collected Icelandic manuscripts and books which are now also in the British Museum (about 120 books and some 30 manuscripts), including copies of the Icelandic version of the Bible, Snori Sturluson's *Edda* and a couple of Icelandic Sagas.[35] Banks had even sent men to Holar in northern Iceland, where the only printing press in the island was located, to buy copies of all books printed there.[36] He collected mineral specimens and examples of Icelandic dress even though he found the Icelandic female costume "certainly not very pleasing to the European eye".[37] He even purchased two Icelandic dogs, appropriately named Hekla and Geysir.[38]

Solander compiled lists of Iceland's flora, fauna and minerals now preserved in the Botany Library of the Natural History Museum. Perhaps Solander's early death explains why they were never published.

So what was the significance of Banks's expedition to Iceland? First: it was the first foreign scientific expedition to explore the island. In 1767 the French government had begun to send expeditions to chart Icelandic waters, but they had spent little time in Iceland itself.[39] For Banks personally the significance of this voyage to the north must have been considerable: as Harold Carter has pointed out, this was "his first and only expedition as indisputable leader".[40] Banks became so fond of the island that he had a map of Iceland engraved on his visiting card, with Mount Hekla of course in pride of place.[41]

From the Icelandic point of view the voyage was a great success. In contemporary Icelandic sources Banks, who was considered a Lord and called Baron Banks, received nothing but praise for his scholarly pursuits, his affability and generosity. The following year the first journal ever to be published in Iceland (albeit in Danish), the *Islandske Maaneds-Tidender*, adorned its front page with an account of the visit of Banks and Solander, commending their generosity and humanity (*Humanitee*).

Finally, as a result of his visit to Iceland and because of his subsequent prominent position in society, Banks was recognized by the British government and individuals alike to be *the* authority on Icelandic affairs, as the following example will demonstrate. In 1801 the northern powers of Europe, including Denmark, combined against England in the League of Armed Neutrality. To combat this very real threat to British naval supremacy the British Cabinet prepared to break up the League. All Danish colonies in the West Indies and India were seized and, as a further retaliatory measure against Denmark, the Pitt administration sought Sir Joseph Banks's opinion on a proposal to annex Iceland. In a lengthy memorandum entitled *Remarks concerning Iceland* and dated 30 January 1801 Banks professed himself in favour of "the conquest of Iceland", a force of 500 soldiers being a sufficient number to "subdue the island without

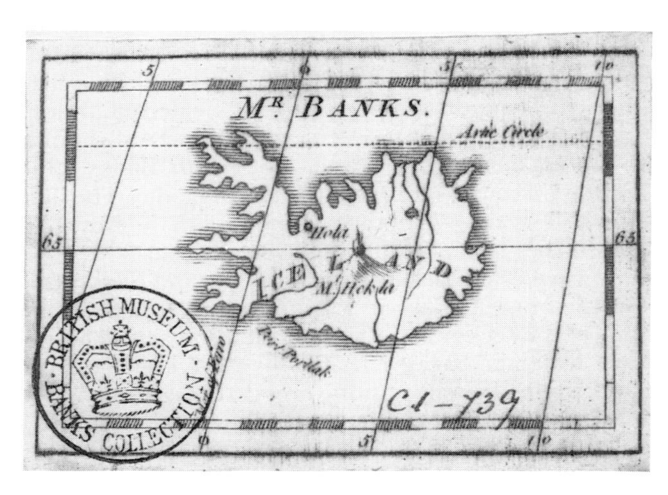

FIG. 5. **Banks's visiting card made after his return from Iceland.**
Reproduced courtesy of the British Library.

striking a blow".[42] During his visit in 1772 Banks had found that "all ranks appeared unhappy and would ... be much rejoiced in a change of masters that promised them any portion of liberty". Thus the proposed annexation would be a benevolent gesture while Britain's gain would be control of the potentially

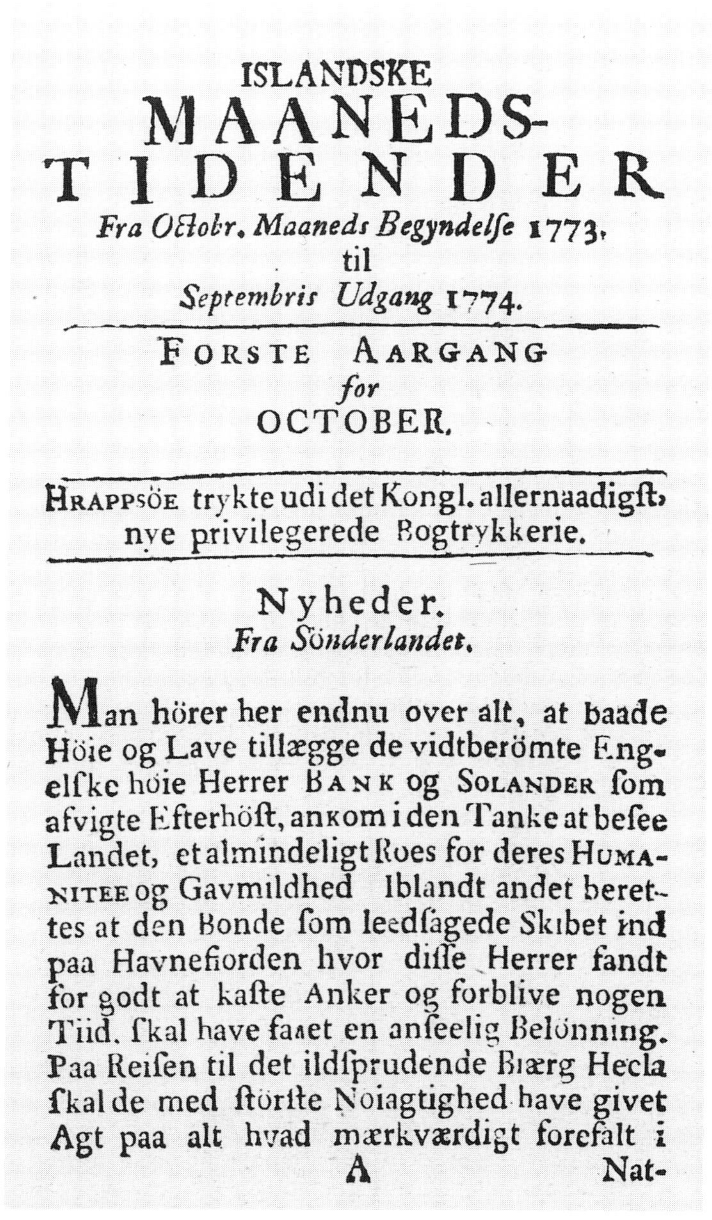

ISLANDSKE
MAANEDS-
TIDENDER
Fra Octobr. Maaneds Begyndelſe 1773,
til
Septembris Udgang 1774.

FORSTE AARGANG
for
OCTOBER.

HRAPPSÖE trykte udi det Kongl. allernaadigſt, nye privilegerede Bogtrykkerie.

Nyheder.
Fra Sönderlandet,

Man hörer her endnu over alt, at baade Höie og Lave tillægge de vidtberömte Engelſke höie Herrer BANK og SOLANDER ſom afvigte Efterhöſt, ankom i den Tanke at beſee Landet, et almindeligt Roes for deres HUMANITEE og Gavmildhed Iblandt andet berettes at den Bonde ſom leedſagede Skibet ind paa Havnefiorden hvor diſſe Herrer fandt for godt at kaſte Anker og forblive nogen Tiid. ſkal have faaet en anſeelig Belönning. Paa Reiſen til det ildſprudende Biærg Hecla ſkal de med ſtörſte Noiagtighed have givet Agt paa alt hvad mærkværdigt forefalt i
A Nat-

FIG. 6. Notice of the visit of Banks and Solander in an Icelandic journal.
Reproduced courtesy of the National Library of Iceland.

important cod fishery. The Danes would be subjected to "a considerable political humiliation in the eyes of Europe" and the possession of Iceland would in the future benefit England's trade, revenues and nautical strength. However, there is no indication that the Pitt administration took any steps to act on Banks's advice and shortly afterwards the League of the Armed Neutrality was dissolved.

In the years following the Iceland expedition Banks continued to take an interest in Icelandic affairs.[(43)] Sir John Thomas Stanley for example, on his return from Iceland in 1789, gave Banks seeds and plants from Iceland, which Banks promised to give to the gardener at Kew. As Stanley wrote in January 1790: "Sir Joseph Banks has been very civil with visits, messages, and invitations, and I dine with him next week".[(44)] Banks initially kept up a correspondence with Ólafur Stephensen, who was to become the only native governor of Iceland ever and who continued to collect and send him manuscripts and scientific specimens.[(45)] But Banks did not really take an active part in Icelandic affairs until 1807 when he received an appeal for help from the son of his old friend Ólafur Stephensen — their friendship providing the link between the two periods of Banks's interest in Iceland.

The Napoleonic Wars 1807-1814[(46)]

In 1807 Denmark, following the British bombardment of Copenhagen, entered into an alliance with Napoleon. Britain and Denmark were thus at war again and the Royal Navy captured many of the Iceland merchant ships on their way to Denmark, bringing them to British ports. Among the passengers travelling on these vessels was Magnús Stephensen, chief justice of Iceland and the leading Icelander of his day. He was the eldest son of Ólafur Stephensen. The younger Stephensen now sought Banks's assistance, reminding him of his friendship with his father and the fact that he himself, at the age of ten, had often had the good luck of seeing Banks when he visited his home. The situation for the Icelanders was desperate; if the navigation between Iceland and Denmark became impossible the Icelanders would face certain starvation, the island becoming "an uninhabited desert".[(47)]

Banks went immediately to the rescue of the Icelanders. As he himself wrote: "The hospitable reception I met with in Iceland made too much impression on me to allow me to be indifferent about anything in which Icelanders are concerned."[(48)] Banks was now a privy councillor and ideally placed to help the Icelanders. In his opinion a British annexation was the only solution to the wartime plight of the Icelanders. He was also convinced of its appropriateness. He wrote: "No one who looks upon the map of Europe can doubt that Iceland is by nature a part of the group of islands called by the ancients *Britannia* and consequently that it ought to be part of the British Empire".[(49)]

As soon as he received the younger Stephensen's letter in October 1807 he sent it to his friend Lord Hawkesbury, the home secretary. Hawkesbury, after discussing the matter with the other ministers, asked Banks to find a way that, in his words, "Iceland could be secured to His Majesty, at least during the continuance of the present war."[(50)] Thus encouraged, Banks set to work planning in minute detail the annexation of Iceland. By December 1807 his plans were ready.

Banks was now unwilling to subject Iceland to what he called "the horrors of conquest". Instead a warship should be sent to Iceland with a negotiator who would meet with the Stephensen family, explaining the benefits of British rule and offering the inhabitants the splendid option of voluntarily becoming British subjects. The Icelanders were simply to transfer their allegiance from Denmark to the United Kingdom by sending their Danish governor to the awaiting warship and accepting a British one in his place. He assured the ministers that in 1772 the inhabitants had been "universally desirous of being placed under the dominion of England".

Expecting perhaps to meet with some hesitation on the part of the ministers, Banks pointed out that once the Icelandic sailors had been "active and adventurous" to the point of discovering America before Columbus, but under the Danes they had degenerated to "their present torpid character". Under British rule, however, they would become "animated, active and zealous", eventually providing plenty of hardy sailors for the Royal Navy.[51]

Though the British government ultimately decided against annexing Iceland, Banks managed to persuade the government to release the Iceland merchantships, that were of course technically enemy vessels. They were, moreover, issued with British licences and permitted to carry on their traditional trade with Denmark in relative safety. The ships were released by the government ostensibly on humanitarian grounds, and this was undoubtedly a factor, certainly this was Banks's motive, while the politicians may have had more pragmatic aims. Napoleon's blockade attempted to close the Continent, one of England's major markets, to British trade. England was to find it most useful to issue licenses to neutral ships (which was to be the acknowledged status of the Iceland ships) to evade the Continental System. As the government was encouraging the search for new markets for British manufactures to compensate for those now closed on mainland Europe, the possibility of an advantageous trade with Iceland may have been an important consideration.[52]

Banks frequently communicated with His Majesty's ministers during the Napoleonic Wars, presenting detailed plans for the conquest of Iceland.[53] He repeatedly urged the government — unsuccessfully — to this course of action. Banks had to admit defeat even though he always believed that a British annexation of Iceland would have been a wise political act, both for Iceland and Great Britain.

The Icelandic Revolution

In 1809 the Icelandic Revolution took place. A British trading expedition, granted a licence to trade in Iceland by the Privy Council, and led by a London soap merchant named Samuel Phelps, seized power in Iceland.[54] The Danish governor was taken prisoner and Iceland was proclaimed an independent country, under the protection of Great Britain, Phelps's interpreter, a Danish adventurer named Jörgen Jörgensen, took up the reins of government as "protector" of Iceland, issuing radical proclamations in the French revolutionary spirit. It was stated that the "poor and common" had the same rights as "the rich and powerful". An Icelandic flag was designed — three white codfish on a blue background — and elected representatives were to meet the following year to draw up a constitution. This colourful episode has been called "the Icelandic Revolution" and Banks has been cast in the role of "a conspirator and revolutionary"[55] and the brain behind the event. Until recently he has been portrayed as such in Icelandic history-books.[56] There is no truth in this. Banks was admittedly involved in encouraging this trading venture, its aim from his point of view being to bring much-needed supplies to Iceland. Banks sent his protégé William Jackson Hooker (later director at Kew) along on the voyage to gather plants,[57] and furnished the party with a letter of introduction to his old friend, Ólafur Stephensen, asking for his assistance in promoting Hooker's researches in the island and recommending Phelps, "a scientific as well as a commercial man" who would he hoped bring back "valuable accounts of many matters, which the shortness of my stay ... prevented me from investigating".[58] He knew the revolutionaries personally — Jörgensen had made Banks's acquaintance in 1806, both having been on voyages of discovery in the Pacific[59] — and he had even, in April 1809, discussed the question of annexation with Phelps and Jörgensen.[60]

Banks, however, was certainly innocent of any planning in this event. The Revolution was simply a spontaneous reaction to the Danish governor's refusal to permit the British soap merchant to trade with the Icelanders. At the time it was the general belief in Iceland that the Revolution had the backing of the British government. There is no evidence to support this, either. In fact it was ended two months later through the intervention of the Royal Navy.

When Banks discovered what had happened, he was deeply shocked, condemning this attempt to involve his innocent Icelanders in what he called "the horrors of a revolution". The Icelandic Revolution was a far cry from the sort of proper peaceful annexation, desired and supported by the natives, that Banks had had in mind. Instead the revolutionaries had actually dared inflict a republican regime on the Icelanders. Banks collected all the information he could on the Revolution, sending accounts to the government of "the atrocities" committed in Iceland.[61] Perhaps Banks' reaction to the affair was so violent, because of a twinge of conscience. After all he had been instrumental in sending these men to the island. But his motives had been humanitarian — to keep Iceland supplied with necessities.

Banks was of the opinion that a repetition of these atrocities — his words — must be avoided at all costs and Anglo-Icelandic relations placed on a firm official footing. In this he was successful. On 7 February 1810 the Privy Council issued an Order-in-Council stating the British government's official policy towards Iceland.[62] Banks was the principal author. Drafts of it, in his unmistakable handwriting, are to be found in the Wisconsin Manuscripts. George III "being moved by compassion for the sufferings of these defenceless people" placed them in a state of neutrality and amity with England. Free trade was established between Iceland and Great Britain and British subjects were forbidden to commit acts of depredation in Iceland. Icelanders, when resident in H.M.'s dominions were to be considered as "stranger friends" and not as alien enemies. The Order acknowledged that the sovereignty of Iceland was still vested in the Crown of Denmark but that Iceland was under the protection of Britain. A British consul was subsequently sent to Iceland, being thoroughly briefed by Banks before his departure.

Throughout the war years Banks was always ready to approach the proper government offices on the Icelanders' behalf. He would never "cease to consider those Icelanders, who may come to England, as friends, who have a right to ask for, and to receive every act of friendship and assistance, I have the power to confer".(63) He bombarded government ministers — e.g. Lord Hawkesbury (later Liverpool), Lord Castlereagh, Earl Bathurst, the Marquis Wellesley — with letters and visits on Icelandic affairs. In all this Banks seems genuinely to have been motivated by Icelandic interests. When for example British fishing interests opposed the licensed Iceland trade and the British government decided to restrict the export of the major Icelandic fish exports, Banks — the Iceland expert — was called in to draft the regulations. This he endeavoured to do as near as possible to the interests of the Icelanders themselves, solicitously enquiring of them the total export figures and fixing the regulations at that level. Banks was thus the architect of the commercial policy adopted by the British government towards Iceland, as well as the political one.(64)

In summary, during the Napoleonic Wars Banks acted as a powerful protector and benefactor of the Icelanders. Without his unstinting and selfless help Iceland would have suffered greatly during this period. Throughout the war years he continuously watched over Iceland's welfare. There is no evidence that Banks ever refused to help an Icelander in need,(65) even to the point of loaning them money from his own pocket.(66) Even the Anglophobe King of Denmark was forced to acknowledge Banks's role as protector of his North Atlantic dependency. At the end of the war a grateful Frederik VI sent a personal letter of thanks to Banks. Banks refused the Order of the Dannebrog, Denmark's highest honour, as a matter of principle,(67) but later accepted three cases of books, including the *Flora Danica* with illustrations of the flowers he missed seeing in bloom when in Iceland in 1772.(68)

Notes

(1) The most important primary source for the Iceland expedition is Banks's journal, unfortunately incomplete (the period from the 6th to 16th of September is missing). For details see H.B. Carter, *Sir Joseph Banks (1743–1820). A guide to biographical and bibliographical sources*, London, 1987, 54. It has been published, with a detailed introduction, by R.A. Rauschenberg 'The Journals of Joseph Banks' Voyage up Great Britain's West Coast to Iceland and to the Orkney Isles July to October 1772', *Proceedings of the American Philosophical Society*, (1973), **117**, (3), 186–226. Secondly, there is the journal of James Roberts, one of Banks's collectors (see Carter, op. cit.), and thirdly Uno von Troil, *Bref rörande en resa til Island MDCCLXXII*, Uppsala, 1777, translated into English as *Letters on Iceland* (the edition quoted from in this paper is the Dublin one of 1780). For more detailed accounts of the Iceland voyage see: H.B. Carter, *Sir Joseph Banks 1743–1820*, London, 1988, 101–112; and two useful modern introductions to von Troil's book by Haraldur Sigurdsson, the editor of the Icelandic edition *Bréf frá Íslandi*, Reykjavík, 1961 and Ejner Fors Bergström, the editor of the 1933 Swedish edition. I would like to express my thanks to Harold B. Carter

for permitting me to use his transcripts of the Banks and Roberts journals in this paper.

The Icelandic scholar Halldór Hermannsson was the first to write on 'Sir Joseph Banks and Iceland', publishing a long essay on the subject in the journal *Islandica* (1928) **18**, 1–99. Though much additional source material has been discovered since then, his account is still valuable and includes a great number of the Banks Iceland letters *in toto.*

(2) von Troil, op. cit. (1), 1.

(3) Banks's Journal, introduction.

(4) von Troil, op. cit. (1), 1–2.

(5) H.B. Carter, *Sir Joseph Banks 1743–1820*, London, 1988, 101–2.

(6) Banks's Journal, introduction.

(7) National Library of Canberra, Banks MS 118.

(8) Hermannsson, op. cit. (1), 4–5; Sigurdsson op. cit. (1), 22–23. Furthermore see: Thorvaldur Thoroddsen, *Landfraedissaga Íslands*, Copenhagen, 1902, vol. 3, 57–58.

(9) Clause Heide to Banks, 25 June 1772, British Library [hereafter BL] Add. MS. 8094, fol. 29–30; Hermannsson, op. cit. (1), 5–6.

(10) Bergström, op. cit. (1), 10–12.

(11) Banks to Hawkesbury, 30 December 1807, Natural History Museum, London, Dawson Turner Collection [hereafter DTC], vol. 17, 78–89

(12) Banks's Journal, 28 August.

(13) Banks to Thomas Falconer, 2 April 1773, Hawley Collection.

(14) Banks's Journal, 29 August to 3 September.

(15) Hermannsson, op. cit. (1), 17–18; G. Meynell and C. Pulvertaft, 'The Hekla lava myth', *Geographical Magazine* (1981), **53**, 433–6.

(16) James Roberts's Journal, 14 September.

(17) Described by Banks in his journal and in his letter to Thomas Falconer of 2 April 1773, Hawley Collection.

(18) Von Troil, op. cit. (1), 10; James Roberts's Journal, 18 September.

(19) Printed in W.J. Hooker, *Journal of a Tour in Iceland in the Summer of 1809*, 2nd. ed. London, 1813, vol. 2, 273–294.

(20) Banks's Journal, 23 September.

(21) James Roberts's Journal, 24 September.

(22) Banks's Journal, 25 September.

(23) So called by James Roberts, 24 September.

(24) Banks's Journal, 6 September.

(25) Von Troil, op. cit. (1), 78; Sveinn Pálsson, *Æfisaga Bjarna Pálssonar*, Akureyri, 1944, 71–2.

(26) J. Espólín, *Íslands Árbaekr í sögu-formi*, Copenhagen, 1854, vol. 9, 3.

(27) P. Langford, *A Polite and Commercial People, England 1727–1783*, Oxford, 1989, 511, (in the series *The New Oxford History of England*).

(28) Banks's Journal, 3 September.

(29) Thorvaldur Thoroddsen, op. cit. (8), vol. 3, 17–56.

(30) Banks's Journal, introduction.

(31) It was reprinted in London in 1808 in the first volume of J. Pinkerton, *A General Collection of the Best and Most Interesting Voyages and Travels.*

(32) G.S. Mackenzie, *Travels in the Island of Iceland during the summer of 1810* Edinburgh, 1811, vii.

(33) Sigurdsson, op. cit. (1), 30.

(34) BL, Add. MS. 15.511–15.512.

(35) Hermannsson, op. cit. (1), 15.

(36) Espólin, op. cit. (26), vol. 9, 3.

(37) Banks's Journal, 6 September.

(38) Espólin, op. cit. (26), vol. 9, 3.

(39) These were Yves Joseph de Kerguelen-Trémarec in 1767–68 and Verdun de la Crenne in 1771–2.

(40) Carter, op. cit. (5), 104.

(41) BL, Department of Prints & Drawing, Sarah Banks's Collection of Visiting Cards, C.1–739.

(42) Remarks concerning Iceland, 30 January 1801, BL, Add. MS. 38356, fol. 39–48; DTC vol. 12; Hermannsson, op. cit. (1), 25–30.

(43) Hermannsson, op. cit. (1), 22 *passim.*

(44) J.H. Adeane, *The Early Married Life of Maria Josepha, Lady Stanley*, London, 1899, 85, 89, 94.

(45) See e.g. letters in BL, Add. MS. 8094.

(46) The subject matter of the second half of this paper is treated more fully in Anna Agnarsdóttir, *Great Britain and Iceland 1800–1820*, unpublished Ph.D. thesis, London School of Economics, 1989 and Carter, op. cit. (5), chap. 22.

(47) Magnús Stephensen to Banks, 17 October 1807, DTC, vol. 17; Hermannsson, op. cit. (1), 33–5.

(48) Banks to Dr. William Wright, 21 November 1807, in A.I. Gans and A. Yedida (eds.), *New Source Material on Sir Joseph Banks and Iceland from the Original Manuscripts in the Sutro Branch California State Library*, San Francisco, 1941, 6.

(49) Banks to Mr. Stephensen of Reikiavick, December 1807, Geographical Society of South Australia, Icelandic Manuscript Collection; DTC, vol. 12. See Agnarsdóttir, op. cit. (46), Appendix 1, 285–7.

(50) Hawkesbury to Banks, 29 November 1807, DTC, vol. 17. Quoted in full by Hermannsson, op. cit. (1), 36.

(51) Banks to Hawkesbury, 30 December 1807, Royal Botanic Gardens, Kew, Archives, Banks's Correspondence vol. 2; DTC, vol. 17.

(52) See Agnarsdóttir, op. cit. (46), chap. 3, 3.

(53) See for example one from 1813, 'Some notes relative to the ancient state of Iceland, drawn up with a view to explain its importance as a fishing station at the present time, with comparative statements relative to Newfoundland', DTC, vol. 17; Hermannsson, op. cit. (1), 75–83.

(54) See Agnarsdóttir, op. cit. (46), chap. 5 and 6; D. McKay, 'Great Britain and Iceland in 1809', *Mariner's Mirror* (London), (1973).

(55) H.P. Briem, '"King" Jörgen Jörgensen', *American Scandinavian Review* (1943), **31**, 124; idem, *Sjálfstaedi Íslands 1809*, Reykjavík, 1936, chap. 48.

(56) Thorkell Jóhannesson, *Saga Íslendinga 1770–1830*, Reykjavík, 1950, 304–326.

(57) Hooker wrote an account 'Detail of the Icelandic Revolution in 1809', in his book, op. cit. (19), vol. 2, appendix A. Subsequently Hooker and Banks frequently corresponded on Iceland. Sir Joseph would not have Hooker "ignorant of anything that has been done relative to our little island" (Banks to Hooker, 1 July 1812, Royal Botanic Gardens, Kew, Archives, Hooker Correspondence, vol. 1).

(58) Banks to Ólafur Stephensen, 28 May 1809, Rigsarkivet in Copenhagen [hereafter RA], Rtk. 373.133, Hermannsson, op. cit. (1) 59.

(59) See Agnarsdóttir, op. cit. (46), 87; see e.g. Jörgensen to Banks, 27 May 1808, and enclosure, Gans and Yedida, op. cit. (48), 72–5.

(60) Banks to Bathurst, 16 April 1809, Natural History Museum, London, Botany Library, Banks Correspondence, fol. 42–3.

(61) See e.g. Banks to Liverpool, 11 December 1809, Public Record Office, [hereafter P.R.O.], F.O. 40/1.

(62) The original in P.R.O., P.C. 1/3901; The *London Gazette*, (1810), 10 Feb.

(63) Banks to Ólafur Stephensen, 28 May 1809, RA, Rtk. 373.133; Hermannsson, op. cit. (1), 59.

(64) See Agnarsdóttir, op. cit. (46), chap. 7.

(65) See e.g. Haldorsen to Banks, 2 December 1807, Gans and Yedida, op. cit. (48), 10–11.

(66) See e.g. Sivertsen to Banks, 31 December 1807, Gans and Yedida, op. cit. (48), 22–3.

(67) Bourke to Rosenkrantz, 26 August 1814, RA, D.f.u.a. 1771–1848, England II, Depecher 1814–15.

(68) Frederik VI to Banks, 17 September 1817, RA, Ges. Ark. London III, Indkomne Skrivelser fra D.f.u.a. 1814–17.

(69) BL, Add. MS. 15.511, no. 13.

(70) See also Carter, op. cit. (5), 111.

(71) BL, Add. MS. 15.511, no. 43.

H.S. Torrens, 'Patronage and problems: Banks and the earth sciences', in *Sir Joseph Banks: a global perspective* (eds. R.E.R. Banks and others), Royal Botanic Gardens, Kew, 1994, 49–75.

5. PATRONAGE AND PROBLEMS: BANKS AND THE EARTH SCIENCES

HUGH S. TORRENS

University of Keele

Abstract

Banks's involvements with the Earth Sciences were on two planes. On the "public" front, his enthusiasm for geology and mineralogy seems never to have been more than lukewarm. As an explorer or promoter of exploration he seems to have taken little interest in these subjects. But as a private individual with estates in Derbyshire and Lincolnshire, he needed geological expertise and his geological patronage was both well informed and crucial to the development of geology.

When geology became institutionalised by the foundation of the Geological Society of London in 1807, Banks's earlier patronage of the newly professional and "practical" mineral surveyors caused serious tensions with the "gentlemanly" geologists of that Society, who marginalised the contributions of the mineral surveyors. The tensions have misled historians into thinking that advances in British geology in the Banksian era were largely "gentlemanly".

A "public" involvement: the years to 1792

"*His knowledge and attention is very much confined to one study, Botany; and his manners are rather coarse and heavy.*" Joseph Farington's diary, 21 January 1796.[1]

If Botany was indeed Banks's passion, there is ample evidence that he also took an astute interest in the earth sciences. But, in dealings with the history of science, we need to separate carefully the *priority* of an individual's achievement against the *influence* of that same individual. This has become particularly necessary in the earth sciences, where debates about the relative contributions made by, for example, Abraham Werner (1749–1817) in Saxony and William Smith (1769–1839) in England have failed to make this vital separation. While there is no evidence that Banks has any priority to geological "discovery" —indeed one is "hard put to come up with any single scientific accomplishment to attach to the name of Joseph Banks"[2] — it is clear that Banks's contribution to the development of geology was both influential and seminal.

Banks took an interest in geological matters while still a student at Oxford as witnessed by his membership of the "Fossil Club at Tittup Hall" near the Headington quarries outside Oxford in 1762.[3] On his southern English tour of 1767 however, Banks had little to say of geological interest; a cave near Cheddar

FIGS. 1 & 2. Views of Staffa drawn on the spot in 1772 by John Cleveley junior.[120]
Reproduced courtesy of the British Library.

which he entered "had nothing at all in it worth the trouble of stopping".[4] But a special visit was made to study Alexander Catcott's famous collection of fossils and minerals at Bristol.[5] The same pattern followed on Banks's later 1767 tour to Wales and then visits to Wednesbury and Dudley.[6] Banks's observations on the fossils of Oxford[7] say little beyond noting the particular fossils occurring there.

Geology also, and perhaps understandably, seems to have occupied little time on the *Endeavour* voyage of 1768–1771, despite the 1768 "hints on minerals and fossils" offered to the expedition by James Douglas (1702–1768), president of the Royal Society.[8] These offered prescient observations on the need to search alluvial materials to learn of the geology of hinterlands. But as an Australian historian has observed, these hints were "overlooked" on the *Endeavour* voyage — "there are no observations, let alone collections, of rocks and minerals to stand with the wealth of material relating to the animal and vegetable kingdoms as souvenirs. Rocks... did travel on *Endeavour*, [but] as ballast".[9]

Banks's next expedition was more geological and it became accidentally more so when, on 12 August 1772, Banks and his Icelandic party reached Staffa and next day saw Fingal's Cave. They were led there by local report that pillars like those of the Giant's Causeway in Ireland occurred. The Irish examples had long been known;[10] and, as far back as 1694, that these columns were made of "*Lapis basaltes*". A "battle of the basalts" about their origins started in mid century. The Frenchman Jean Etienne Guettard (1715–1786) demonstrated the existence of extinct volcanoes in the Auvergne in 1751.[11] This became a "Vulcanist" explanation. Rev. Richard Pococke (1704–1765) instead deduced that these pillars were formed by repeated precipitations from water or mud to give a "Neptunist" view.[12]

In May 1771 Nicolas Desmarest (1725–1815) read to the *Académie des Sciences* in Paris his famous paper on polygonal basalt columns. He had suspected they were volcanic in origin in 1763 and his conclusion that they were, was announced in 1765 and published in 1768.[13] This reached the Royal Society in 1770 when R.E. Raspe confirmed that similar basalt columns near Cassel in Germany were volcanic in origin.[14] This, and oral reports of the 1771 paper from Paris, seem to have inspired Banksian interest in such phenomena and aided the choice of Iceland for his 1772 expedition.

Banks reported in some detail on Staffa. He noted the pillars were "of lava-like material of a coarse kind of basaltes".[15] His detailed description was not published in 1771[16] nor was it, as in an earlier claim, "a phenomenon till then unobserved by naturalists".[17] Banks does however deserve credit for having first brought the wonders of Staffa to the proper attention of "philosophers". Figures 1 & 2 are of Staffa drawings taken on the spot in 1772 by John Cleveley junior.[18] Joseph Farington's observation that "*accuracy* of drawing seems to be a principle recommendation to Sir Joseph" — diary entry 12 December 1793[19] — seems entirely confirmed. It is sad that the value of the party's detailed observations on volcanic columnar basalts was dissipated by the "volcanic illusion" which Daubeny[20] was still able to discuss so eloquently 50 years later.

The expedition to Iceland had a true mineralogical dimension. Uno von Troil (1746–1803), the Swedish priest, was a member and took a real interest in such matters. It was his account of Iceland that was published, not Banks's; supplemented with remarks by Torbern Bergman (1735–1784) who discussed volcanic effects in Iceland and the origins of basaltic pillars. Bergman took these to be an effect of moist substances which had burst while drying to cause contractions.[21] Banks again used rocks merely as ballast and many of his Icelandic lava specimens were supporting moss gardens at Kew by 1784![22]

Between 1781 and 1820, i.e. during most of Banks's presidency of the Royal Society, Carter noted that only 3.6% of papers published concerned geology. Even taking a broader view of Earth Sciences, only up to 7% can be so considered.[23] Banks's personal involvement with such matters up to 1792 was equally episodic.

Banks's involvements with Australian exploration from 1788 were legion but again one notes the strange lack of interest in things mineral: on *Endeavour* [Solander's] "attention was reserved.... for plants and animals".[24] There was a similar lack of any real interest in things mineral on the HMS *Investigator* expedition. Banks instructed Robert Brown in 1801 that "Geology and Mineralogy must be considered by you as subsidiary pursuits and you will be required to do in them no more than is compatible with a full attention to Botany, Entomology, Ornothology etc".[25] Likewise when mineralogists went out to South America (John Mawe in 1804) and Australia (Robert Townson in 1806)[26] there is little evidence of any Banksian enthusiasm for their activities, although Banks was involved with both projects. In a *laissez-faire* society they went out as *individuals* in the eyes of Banks. All this, for two recent writers "betrays the [then] limits of Banksian vision in science".[27] This vision was soon to change.

A "private" involvement: Banks and mineralogy, 1792 to 1801

"*The Gregory Mine into which I twice descended actually passes under* [*Sir Joseph's*] *house and several veins of lead are exposed to view in his gardens. Overton Hall therefore may be called a Mining Villa and the fit residence of a mineralogist.*"[28]

In 1792 Banks's private attitude to geology was forced to change. He inherited estates at Overton near Ashover,[29] which had come into the Banks family through inter-marriage with the Hodgkinsons.[30] Up to this time Banks had had limited contact with geology. London long after 1792 was still thought a place where the study of geology was impossible, since flint was "almost the only branch... that can be practically studied" there.[31] The Banks estates at Revesby in Lincolnshire were similarly thought to be in an "uninteresting" region for geology. But Derbyshire was different. Lead had been mined at Ashover since at least the seventeenth century[32] and the Overton mines and estate revenues so intertwined that Banks's income depended on them. One historian has claimed that for Banks "the mineral kingdom was no more than a source of income".[33] If this was ever entirely true, it is surely most true of this decade.

The Gregory Mine at Ashover had been the county's richest and most famous mine. The "clear profit at Gregory Mine to the Proprietors [of whom Banks was one] from the year 1758 till Xmas 1806 ... was £101,535".[34] But between 1783 and 1788 profits dropped by nearly 60% and by 1792 the mines at Overton were becoming a liability. John Milnes noted "from Lady Day 1790 till Xmas 1803 there was a loss paid as per quarterly reckonings of £23,398"[35] One of Banks's first actions on inheriting was to draw up "memorandums made at

Overton — March 1793."[36] [John] Nuttall [fl.1765–fl.1830] the local Matlock land surveyor was to make a map of the parish, to survey Banks's lands at Alton "where coal is supposed to be" and then to suggest how such coal might be proved geologically. Clearly Banks was being forced to widen his mineral horizons with the help of his local mineral agent William Milnes II (c. 1757–1814).

A letter Sir Charles Blagden wrote to Banks on 8 September 1794[37] provides further evidence of Banks's new search for better mineral information. It notes that "the Duchess of Devonshire[38] said she had recommended to you Mr Watson, who makes the tablets representing the strata of Derbyshire and she speaks in very high terms of the talents and mineralogical skill of Mr Barker and Mr French, two young gentlemen of that county". Banks had already made contact with White Watson (1760–1835), whose commonplace book[39] notes that Banks first called on him on 17 August 1794 when they had breakfasted together with "Mr Milnes of Ashover". Thomas Barker (1767–1816), lead merchant of Ashford Hall, Bakewell, had been a pupil of Abraham Werner in Freiberg in Saxony in 1790 and 1791.[40] Dr Richard Forester French (1771–1843) of Derby had helped Erasmus Darwin with some of his experiments as early as 1787.[41] French had, in 1792, just graduated M.B. He changed his name to Forester in 1797.[42] He played little further part in advancing Derbyshire geology, although elected an Honorary Member of the Geological Society of London in 1809.[43]

Charles Hatchett (1765–1847), a further visitor to Overton in June 1796, provided vital information of what was known of Ashover geology at that time. His diary[44] records the exposed downward stratigraphic sequence of grit, shale, limestone overlying toadstone in a sketch (Figure 3). This shows that these rocks were then known to occur in a domed structure which had been split by the river Amber. Concepts of such domes — later to be named anticlines — were already familiar, if unnamed, by 1760.[45] Denudation by subsequent erosion was equally clearly known. Hatchett also confirmed the financially unpromising nature of some of the local lead mines in 1796.

The following year saw the first recorded visit to Derbyshire by the man who, with Banks's vital patronage, subsequently helped revolutionize the study of both British and Derbyshire geology: John Farey senior (1766–1826). Farey noted that he first visited Derbyshire in 1797 when examining bogs near Buxton.[46] In June 1798 Farey, who was then estate steward to the Duke of Bedford at Woburn, first met Banks. Farey, who at that time had little real interest in geology, explained to Banks "the extensive system of drainage and irrigation" he was supervising on the Duke's estates.[47] In 1800 Farey first heard of William Smith (1769–1839) the geologist[48] and met him in October 1801, when the relationship which led to Farey being called "Smith's Boswell" commenced.

In the same year one of the last of the old school of strictly mineralogical geologists visited Banks at Overton; the "Neptunian" John Hawkins (?1757–1841) of Cornwall who had also studied at Freiberg under Werner, in 1786 and again in 1793. He wrote to fellow mineralogist Philip Rashleigh (1729–1811) from Buxton on 14 September giving his impressions of Overton. He had met there the "Plutonian" Scotsman Dr Thomas Hope (1766–1844) but they had been unable to resolve their differences about the origins of the toadstone exposed at Ashover, which John Whitehurst had claimed in 1778 was volcanic in

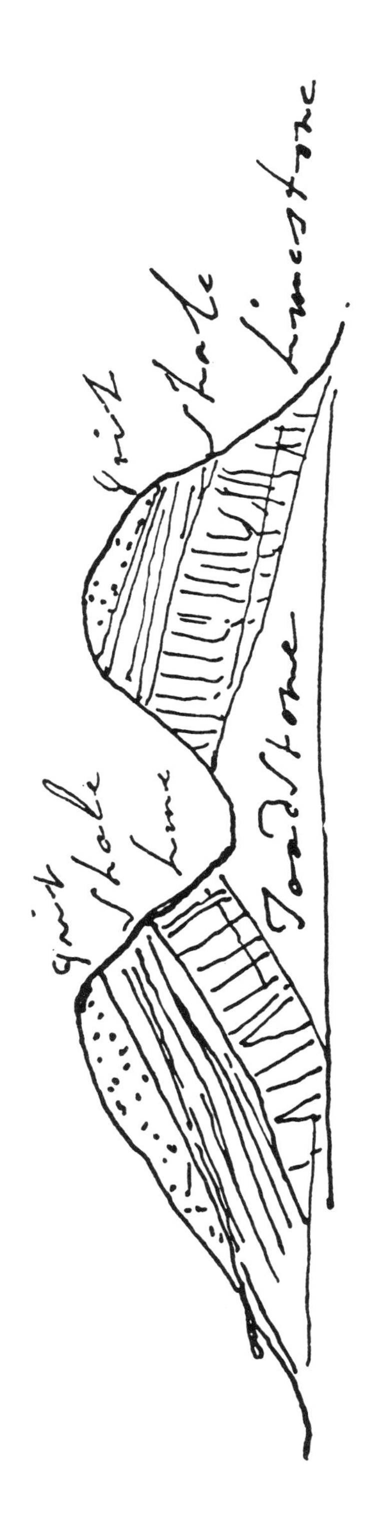

FIG. 3. Sketch of the stratification at Ashover by Charles Hatchett. 1796.[121]

origin. But Hawkins *had* been able to study the relations of the lead veins to both the enclosing limestone and the toadstones. He saw how the high dips of rocks made life difficult for mineral adventurers at Ashover when seeking lead below the overlying shale cover. This was one of the reasons Hawkins thought "the state of mining industry here is much on the decline. The produce of the Derbyshire mines being reduced to one half of what it was ten years ago". In attempts to improve mining output John Milnes (c. 1770–1838), William's brother, prepared a detailed "section of the strata at Gregory's Mine" in June 1801 which he improved up to 1806(49) and, in about 1803, a map of veins and mines in the parish of Ashover.(50)

Banks's private and "public" involvements in geology now came together, since it was John Hawkins who provided the Geological (or as he called them "Mineral Geographical") Instructions which the crew of H.M.S. *Investigator* took to Australian waters in 1801.(51) The intentions of this expedition were chiefly hydrographic and botanical. So the "Instructions" started off in defensive mood, "it would be wrong not to pay some attention to Mineralogy" but conceded that "this department of natural History promises fewer discoveries than the two others [animals and plants] which are ... the objects of the expedition". The influence of Banks is easy to see. It may have been during Hawkins's September 1800 visit to Overton that these instructions were drafted. If so, Banks may have played a part in writing them. John Mawe (1766–1829), the London mineral dealer, was also involved in proffering advice to the *Investigator* expedition.(52)

In order to pay lip service to the need for *some* mineral expertise, Banks had been instrumental in renewing negotiations for a "practical miner" to join this expedition in January 1801. John Allen (born Ashover 1775) was the man chosen(53) but he had only a subsidiary role, working under the direction of naturalists in collecting rocks. He was not initially expected to make trials below the surface, although Banks had changed his mind on this by June 1801. The limitations of sending such a man, however capable a miner, were revealed by the trained mineralogists attached to the French *Baudin* expedition the following year.(54)

A private involvement: Banks and stratigraphy, 1802 to 1811

While efforts were made to save the Ashover mining industry from the terminal decline which finally struck in October 1803, Banks, as we have seen, had come into contact with John Farey (Figure 4) and his "master" William Smith. The chronological connection between this decline and Banks's espousal of their "new technology" of mineral surveying is so close that it seems one was a direct consequence of the other. Proof in the face of the wide or total dispersal of manuscripts of all three protagonists is more difficult. But between 1800 and 1804 we note a total swing in Banks's allegiances from Wernerian "mineral geographers", like Hawkins, to native "mineral surveyors". The close connections between mineral wealth and industrial opportunities meant that Banks could hope for some real financial rewards.

Relations between all three men have been described by Eyles (1985). Banks first met Farey in 1798, while Banks and Smith first met in the summer of 1801; when Banks "favoured Smith with an interview, and from this time till his death remained a steady friend and a liberal patron of his labours".[55] Smith had just published his abortive but widely circulated *Prospectus* in June 1801. The Royal Society copy is annotated by Banks "sent my name to Debret". But Smith's plans to publish, and Sir Joseph's hopes to subscribe, were set back when John Debrett (died 1822), Smith's intended publisher, was declared bankrupt in October 1801. Smith was now entangled in the financial problems which beset him for the next 30 years.

FIG. 4. Silhouette of John Farey, probably drawn by White Watson, in 1807.[122]

FIG. 5. Portrait of Sir Joseph Banks in 1802 by Rembrandt Peale (1778–1860).[123] Reproduced courtesy of the Library, The Academy of Natural Sciences of Philadelphia.

Banks's view of the significance of Smith's discoveries was greatly stimulated by Farey, who on 11 February 1802 wrote Banks a significant letter urging the *importance* of what Smith had discovered.[56] Figure 5 shows a newly located portrait of Banks from this same year. Farey had just returned from a triangular geological "tour of the strata" through Buckinghamshire and Bedfordshire early in 1802, with Smith and Benjamin Bevan (1773–1833), another of Smith's important pupils. This tour had demonstrated to Farey the importance of Smith's discoveries.[57]

Smith had made two advances of great significance for geology. The first was to document the *sequential* order of British rocks. The scale of Smith's achievement here, in building on what little was previously known, can be best seen from the chart published by Challinor.[58] The second was Smith's discovery of the means to *identify* individual strata with that sequence by the organised fossils found in, and often characteristic of, particular rock units.

Banks's response to Farey's letter was characteristically helpful. In May 1802 Smith noted in a letter to his relative Samuel Collett[59] "I have been obliged to have recourse to an uninterrupted pursuit of my subject [land and mineral surveying] with all my plans and papers together in this place [Bath] and shall call on Sir J. Banks in London to settle abt bringing my Papers before the Publick in a much better form than if they had appeared last year. I am now confident they are correct & my map begins to be a very interesting History of the Country".[60]

Banks and Smith attended the annual sheep-shearing meeting at Woburn in June 1802 when "Smith ... exhibited his map now in very considerable forwardness, of the strata of different earths, stones, coals & which constitute the soil of this island ... he was particularly noticed by Sir Joseph Banks".[61] Smith next wrote to Banks at Revesby in October 1802 outlining what little progress he had made. On 21 April 1803 Smith showed Banks "my plan of strata" in London.[62]

In June 1804 the annual Woburn sheep-shearing was again held.[63] Banks, realising the financial plight Smith was in, subscribed £50 as a starting subscription towards the Geological map Smith now hoped to publish.[64] The undated *Proposals* for publishing this, which Smith then circulated, survive in the Bedford papers[65], in Smith's hand. At the Smithfield Cattle Show in December 1804 "Mr Smith... shewed his map of the Strata of England, which is preparing for Sir Joseph Banks, in whose library it will soon be exhibited for the information of the curious".[66] On 29 January 1805 Smith again saw Banks in London according to his diary. He was clearly now having considerable problems struggling to earn a living while finding time to complete his map. His problems also involved adjusting to literary life.[67]

John Farey realised how acute these problems were, and so on 21 May 1806 wrote to the *Philosophical Magazine* to publicise Smith's work and the "want of sufficient public encouragement" that Smith had received. Farey now published "the rules... for ascertaining the relative position... of each distinct stratum, however thin, with regard to those above and below it in the series [of British rocks]" which Smith had discovered. The first was "by the knowledge of its relative position with other known strata in its vicinity" [what geologists would call superposition]. The second "by the peculiar organised remains imbedded in it and not to be found in the adjoining strata". The third, to be used when all else

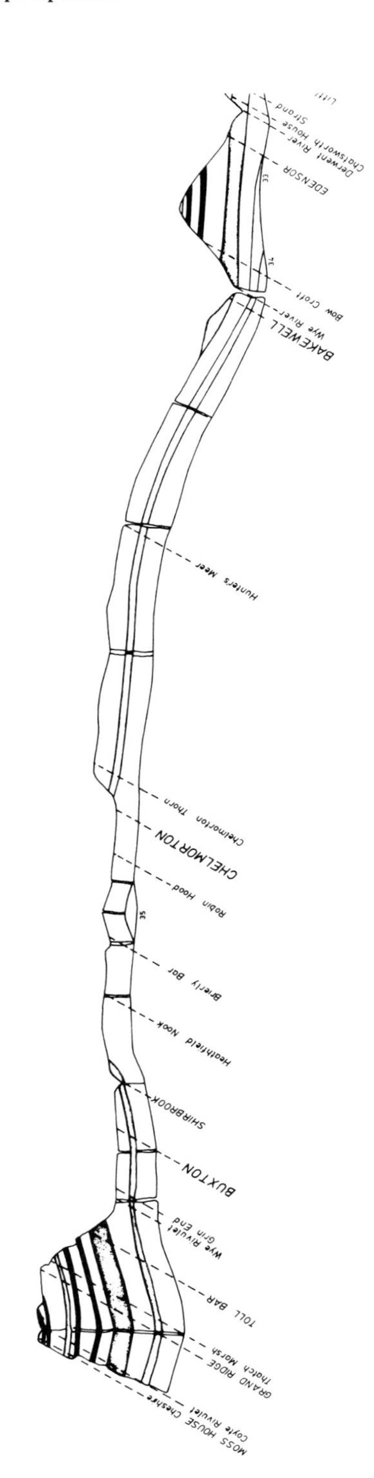

A Section of the STRATA of DERBYSHIRE through BUXTON and BOLSOVER; By After Mr. JOHN FAREY'S Survey of

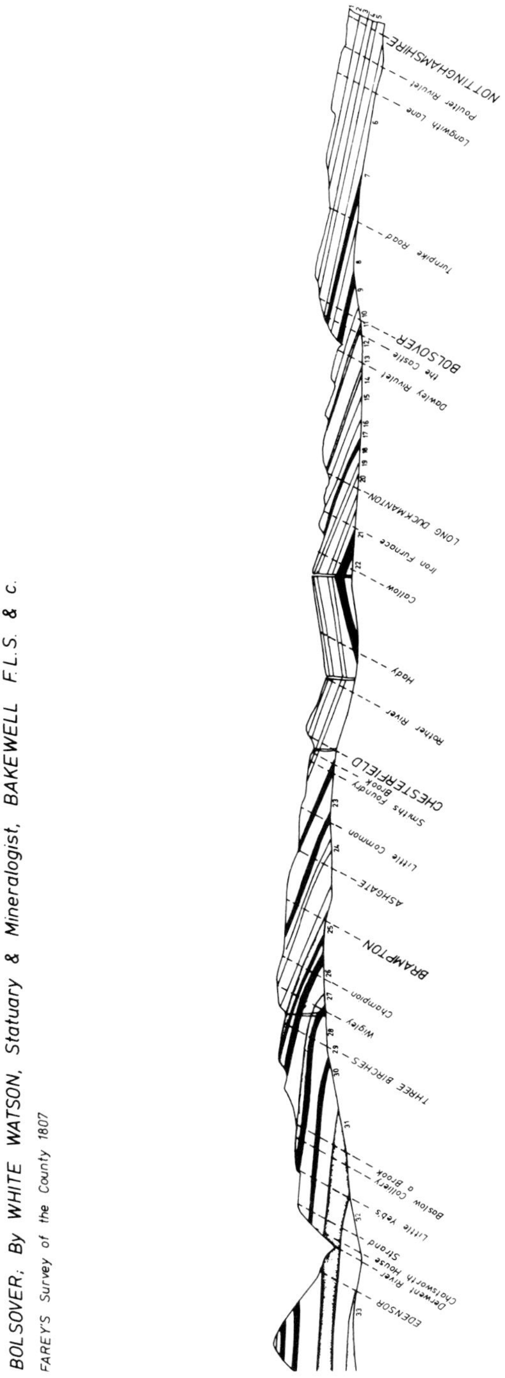

FIGS. 6 & 7. Farey's first geological section across Derbyshire, which he allowed White Watson to copy in November 1807.(124)

had failed, relied on the unreliable; "the peculiar nature and properties of the matter composing the stratum itself" [its mere lithology].[68] Ominously Farey wrote to Smith only two days later to note Banks's wish that Smith's work should be completed.[69] But Smith continued in financial difficulties; as he noted to Richard Crawshay on 27 August 1807 the "subscription which Banks had opened in 1804 had not been so readily encouraged".[70]

It was against this background that Banks turned to Smith's better placed and London-based pupil, John Farey with his personal patronage. Banks's first commission was in July/August 1806 when he asked Farey (whose expenses he met) to undertake a "section of the earth from London to Brighton". This was finished and dedicated to Banks early in 1807,[71] but never published in either's lifetime. Banks then asked Farey to make some calculations for the use of his lead mining agent in Derbyshire", as his introduction to Derbyshire. Things were complicated when, later in 1807, Richard Phillips (1767–1840) introduced Farey to Sir John Sinclair (1754–1835), President of the Board of Agriculture. Two parallel projects, both involving Farey and Banks in Derbyshire, now developed.

The first started in August 1807 as a geological survey of Derbyshire "at the instance of [Banks] worthy President of the Royal Society in order to examine minutely its Stratification and Mineral Treasures". At the beginning of September 1807 Farey joined Banks on his annual visit to Overton for a week. The first product of this work was a section of Strata across Derbyshire which Farey allowed White Watson to copy in November 1807 (Figures 6 & 7). Word soon spread about Farey's work for Banks. A.B. Lambert (1761–1842) wrote to William Cunnington (1754–1810) on 23 December 1807 noting; "the last time I was in Soho Square Sir Joseph showed me a curious and interesting drawing that has just been finished by a pupil of your friend Smith, of all the Strata of Derbyshire on Smith's new principles and Sir Jos. was much delighted with it".[72] The section was then sent in by Banks to be exhibited at the second meeting of the new Geological Society of London (at which Banks was elected a member) on 1 January 1808.[73]

Farey was soon also busy on a second and more detailed stratigraphical section right across England from the Lincolnshire coast through Banks's estates at Revesby to end at Overton. Farey moved from Overton to start work at the eastern end of this section, examining the strata near Boston and Revesby in October 1807.[74] Here Farey witnessed another of the many abortive attempts to find coal in strata which merely and superficially resembled coal-bearing Coal Measures and which misled the "adventurers" concerned. Such trials were frequent before the work of the newly professional mineral surveyors like Smith and Farey had established a scientific basis for mineral prospecting. The true geological horizon of this trial near Raithby was identified, nearly correctly, by Farey using "the proper extraneous fossils" he found in these strata, as Smith's Clunch Clay; merely one of many rock units much higher in the British sequence above the Coal Measures which "resembled" coal-bearing beds.[75]

This long and detailed stratigraphic cross section, from the Lincolnshire coast to Overton, Farey finished on 17 February 1808 and dedicated to Banks. On it Farey styled himself Smith's pupil. The section was 3 metres long.[76] Four contemporary copies survive.[77] It was exhibited at Lord Somerville's Spring Cattle Show in London on 29 February 1808.[78] Clearly it was widely distributed, despite not being published. In the following month, members of the Geological Society of London visited Smith, and Farey, who had been present on this visit, opened his correspondence with G.B. Greenough — the president.[79] All parties were hoping to advance the science of geology.

A third fruit of Banks's Derbyshire commission to Farey was a coloured geological map of the NW portion of Derbyshire dated September 1808 of which the original MS also survives in the Sutro Library,[80] with a sketch explaining faults in the "great Limestone District" in Derbyshire.[81] Farey's now improved "section of the Strata of the Great Mineral Limestone District and its bordering Strata in Derbyshire" dated September 1808 also survives.[82] None of these were ever published, but the "mineralogical map by Mr Farey" was exhibited to the Geological Society by the President who had received a copy from Banks on 2 December 1808. His improved September 1808 Derbyshire "mineral section" was also exhibited to the Geological Society at Banks's request on 3 February 1809. But Banks resigned from the Society the following month.

The second collateral Derbyshire project Farey had agreed to undertake for Sir John Sinclair. He "had previous to my [Farey's] setting out on that [Banks] Survey engaged me to collect at the same time, the necessary facts and particulars for a *Report to the Board* [of Agriculture], on the Agricultural and Rural Concerns of the County".[83] The details emerge from letters.[84] The first is a copy of Sinclair's reply to Farey, accepting Farey's terms; the second is Sinclair's explanatory letter to Banks. Sinclair now agreed that Farey enter into "greater details concerning minerals and soil in Derbyshire than had been normal with earlier *County Agriculture Reports*". Sinclair reported to Farey that Banks "seems much impressed" with the idea that Farey should be commissioned to do the Derbyshire *Report* and it was now agreed that Farey be commissioned to do this, for £300 "in full of all Farey's demands". Sinclair sent a copy to Banks with a covering letter noting "if the Board says the allowance is too much, I shall shelter myself under the *Banks* of literature". Banks merely annotated this letter on receipt "not a syllable in my letter has a reference to the increased price offered to Mr Farey" and continues "I think he will deserve it, if he is able to give a distinct account of the Stratification of Derbyshire but whether he is or not remains to be proved". It was impressive that Farey was so soon able to answer Banks's fears.

The Survey of Derbyshire, which the Board of Agriculture had started to take off Banks's hands, now commenced in earnest, and from the late summer of 1807 for two years Farey was busy with his Geological and Agricultural Survey. By a new agreement with Sinclair in April 1810 it was agreed that the Board of Agriculture should purchase, for £150, much of the research which Banks had originally commissioned for incorporation into the *Derbyshire Report* which would now include a general summary of the stratification and mineral concerns of the county as the first volume of the *Report.*[85] This was to have become the first in an intended series of *County Mineral Reports* to accompany the *Agricultural Reports.* By 20 November 1810 the first chapter of Farey's *Derbyshire Report* —covering geology — was completed. It comprised the whole of one volume; an octavo which gave a pioneering survey of British Strata, published in June 1811 (see Figure 8). It broke new ground with a long discussion on "terrestrial stratification", announced Smith's major results and listed the 21 strata down to the Red Marl — the highest unit found in Derbyshire

GENERAL VIEW

OF THE

AGRICULTURE

AND

MINERALS

OF

DERBYSHIRE;

WITH

OBSERVATIONS ON THE MEANS OF THEIR IMPROVEMENT.

DRAWN UP FOR THE CONSIDERATION OF

THE BOARD OF AGRICULTURE

AND INTERNAL IMPROVEMENT.

VOL. I.

CONTAINING A FULL ACCOUNT OF

The Surface, Hills, Valleys, Rivers, Rocks, Caverns, Strata, Soils, Minerals, Mines, Collieries, Mining Processes, &c. &c.

Together with some Account of the recent Discoveries respecting the Stratification of England; and a Theory of Faults and Denudated Strata, applicable to Mineral Surveying and Mining.

ILLUSTRATED BY FIVE COLOURED MAPS, AND SECTIONS OF STRATA.

BY JOHN FAREY, SEN.
MINERAL SURVEYOR,
OF UPPER CROWN STREET, WESTMINSTER.

LONDON:

PRINTED BY B. McMILLAN, BOW STREET, COVENT GARDEN:
SOLD BY G. AND W. NICOL, BOOKSELLERS TO HIS MAJESTY, PALL-MALL; SHERWOOD, NEELY AND JONES, PATERNOSTER-ROW; DRURY, DERBY; BRADLEY, CHESTERFIELD; AND TODD, SHEFFIELD.

1811.

FIG. 8. Title page of Farey's 1811 Mineral *Report on Derbyshire*, volume 1.[125]

— which had been elucidated. The strata found below in Derbyshire were then sequentially discussed and a significant section on Faults with coloured diagrams included.[86] Finally Farey "sincerely hoped that such a desire may speedily be excited, for the publication of... [Smith's] valuable Maps and Papers, and descriptions of his Fossils, illustrative of the British Strata as would induce him to lay by his professional engagements, in order to publish them".[87] This hope was sadly to be dashed; although Farey's complete *Report* was published in three volumes over 1811, 1813 & 1817 totalling over 1900 pages for a payment to Farey of £450. This was despite the bankruptcy of the publisher in November 1810 and Sinclair's financial crisis in 1815. Farey's *Derbyshire Report* has rightly been called "magnificent" by Riden.[88]

A final act of this decade was the reading to the Royal Society in March 1811 of Farey's letter to Banks dated 31 January. It described "the great Derbyshire Denudation", and demonstrated the geological structures Farey had uncovered in Derbyshire.[89] But it twisted a knife in the wounded relationship between the Royal Society and the fledgling Geological Society founded in 1807. This wound was made when Banks resigned from the Geological Society in March 1809.[90] It continued to fester for many years after Banks's death.

The Final years: Banks in decline, 1812-1820

This period was overshadowed by Banks's ill health[91] which allowed him fewer opportunities to support the work of "his" mineral surveyors. That they needed help is clear. A letter from Farey to G.B. Greenough of 16 September 1810[92] shows that Farey was already at work on an intended paper for volume 1 of the *Transactions of the Geological Society*. This was to be a modified version of the original Banksian commission; a "minute Mineral Survey of the parish of Ashover and its environs".[93] A second letter of 27 May between the same parties shows that the Ashover map and section could not be read before the Society until the explanatory memoir had been written, which Farey was then busy writing. Farey and his son William had restarted this project in August 1811,[94] after publication of the *Derbyshire Report*.

By 2 December 1812 the large mineral map of Ashover was ready and it was shown to both Banks and Greenough. By 25 December 1812 a diminished version was also ready and good progress had been made with the memoir and the large map and section with the help in Ashover of William Milnes III (1785–1866). The memoir was delayed in December 1812 until Farey had seen the Cuvier & Brongniart memoir which was "not yet in Sir Joseph's library". By 1 January 1813 all was finished, apart from a detailed description of the 24 strata exposed in the Ashover area. The tragedy is that this Ashover work was never published and has nearly all been lost, like most of Farey's many other manuscripts. The only fragment to survive is the reduced map reproduced here as Figure 9. It is a miracle of detailed geological mapping for its date, December 1812, on a scale of 1½ inches to a mile.[95] 24 separate strata are separately coloured geologically.

The Geological Society Council, however, took a less favourable view of this work, as did a historian of the Royal Society.[96] After noting that Farey had not been previously acquainted with Wernerian geognosy and that his work was

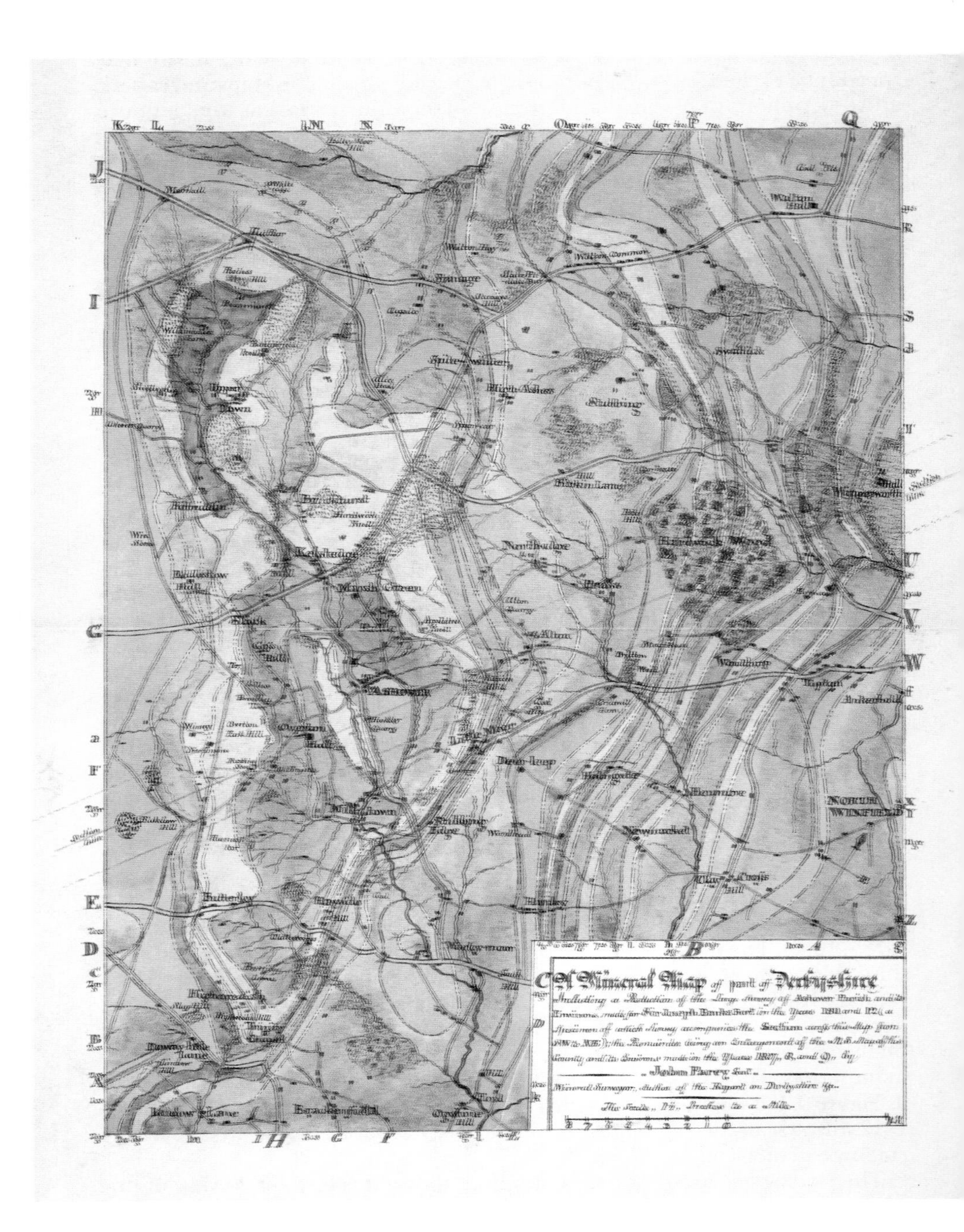
A Mineral Map of part of Derbyshire
John Farey

independent, Thomson noted that Farey's stratigraphic observations "certainly do him credit as an observer but are unfortunately too circumstantially minute for common use. General principles stripped of all useless details are much more attractive and much more easily understood and remembered than a multitude of minute observations". The scale of the problem Farey faced in getting this paper published is revealed by his letter to Greenough of 23 January 1813 which refers to further sections of the text having been removed by officers of the Society. It was this version which was formally read in February.[97] By 8 April 1813 the paper could only be published if it was further shortened by the Society, and Farey pleaded with Greenough that his three years work should not be so badly mutilated. If it was to be, Farey asked, on behalf of himself and his patron Banks, that it be returned as soon as possible. But by 29/30 May 1813 a four to one abridgement had been made. Farey complained that it meant this was a new paper; "a stranger to me and my subject". By June 1813 all negotiations had failed and Farey asked for, and eventually was granted, the return of his paper. Farey's correspondence with Greenough now ended. The referee who had been given the job of "seeing it through the press" was Henry Warburton (1784–1858). One wonders if the real reason it was never published was a vendetta against Banks for his resignation in 1809?

A letter of 7 July 1813 from Farey to Sowerby shows he now hoped to publish his Ashover paper "rejected thro' the intrigues of a few individuals by the Geological Society", with another Report he had made on the Alum Shales of Yorkshire, as the first volume of an intended new series of *Mineral Reports* "if I can meet with any bookseller willing to undertake it". But Farey could not find a publisher and in one of several outspoken attacks wrote of those at the Geological Society; "hostile to the cause of real Geological Science, as to those persons engaged in practical and untheoretical investigations of the facts of British Stratification", who had aborted his paper. His failure to achieve separate publication showed it "could not be published at all in this country", such was the lack of interest in supporting such projects.[98] With Banks's ill health, interest in such projects was limited, especially when vetoed by the gentlemen geologists of the Geological Society or plagiarised by others (like White Watson in Derbyshire).[99]

FIG. 9. Farey's remarkable 'mineral map' of Ashover,[126] finished in December 1812. The original scale of 1½ inches to 1 mile has been reduced for this publication (see scale bar on map). The sequence of all 24 strata mapped by Farey (many are identified round the periphery, where *cs* = clay-shale, *gr* = Grit) can be identified as follows from below.

1) *red-brown* [in centre of the Ashover anticline[127]] = Toadstone [Tuff]; 2) *green* = [Carboniferous Limestone] (with Overton Hall on its outcrop); 3) *brown* = [Head & Boulder Clay]; 4) *yellow* = [Ashover Grit]; 5) *grey* = 1st clay-shale; 6) *lighter brown* = 2nd [Chatsworth] Grit; 7) *grey* = 2nd clay-shale; 8) *pink* = 3rd [Crawshaw Sandstone] Grit; 9) *grey* = 3rd clay-shale (note the coal pits marked at Alton); 10) *red* = 4th Grit; 11) *grey* = 4th clay-shale; 12) *yellow* = 5th Grit; 13) *grey* = 5th clay-shale; 14) *brown* = 6th Grit; 15) *grey* = 6th clay-shale; 16) *pink* = 7th Grit; 17) *grey* = 7th clay-shale; 18) *red* = 8th Grit; 19) *grey* = 8th clay-shale; 20) *yellow* = 9th Grit; 21) *grey* = 9th clay-shale; 22) *brown* = 10th Grit (with Wingerworth Hall on its outcrop); 23) *grey* = 10th clay-shale; 24) *pink* = 11th Grit.

[Beds 9 to 24 are now classed as Lower Coal Measures; a series of interbedded sandstones and shales, often of more limited lateral extent than Farey realised in 1812]

The lack of publication outlets for mineral surveyors must be taken into consideration by historians. The debt of later members of the Geological Society to the mineral surveyors has simply not been acknowledged. The *eight* different charts of the "Order of Superposition of Strata in the British Islands" which William Buckland produced between 1814 and 1821 are a prime example. One historian has claimed "it was precisely in seeking to transcend local descriptive stratigraphy that the intelligentsia distinguished themselves, ... [this] becoming one of the hallmarks of the gentlemanly geologist distinguishing him from the practical geological surveyor".(100) This ignores the fact that the latter had laid the groundwork for descriptive stratigraphy in England! Farey's 1812 map reproduced here shows at last the abilities of such people and the remarkable detail of which they, alone then in Europe, were capable.

Banks put his finger on the true problem in September 1811 when writing to the French geologist Faujas de St Fond; "Geology becomes more & more a fashion. I hope we shall before long advance somewhat the Limits of that Science. We have now some Practical men well versd in stratification who undertake to examine the subterraneous Geography of Gentlemens Estates in order to discover the Fossils likely to be useful for Manure for fuel etc... if employment begins to be given to these people the Consequence must be a rapid improvement if the Labour in this great work can find means".(101) But no further *County Mineral Reports* were ever published by the Board of Agriculture, which found itself in serious difficulties because of the recession at the end of the Napoleonic Wars.(102) This badly affected business openings for the mineral surveyors and meant that Banks's aspirations for them remained unfulfilled.

The tensions between Banks and Greenough also need examination. Banks who "determined never to enter Parliament"(103) was landed and above all utilitarian. Greenough, first President of the Geological Society, on the other hand was Member of Parliament for the rotten borough of Gatton (with only one voter), and *nouveau-riche* and polite. Greenough's wealth had come not from land but from patent medicines!

Such tensions have left a fertile field for revisionist historians who have not ensured that all sides are equally represented. One group published fully, the other did not. Some of the divisions are across that most destructive of English barriers, Class; with Gentlemen against Players; "Practical Men" versus "Cabinet Philosophers"; or "the Establishment" against dissent. Jack Morrell claims that there has been too much "emphasis on gentlemanliness in Victorian English science"(104) and the situation regarding Banksian times is the same.

The years 1812–1820 were tragic ones for the mineral surveyors. Smith was given a further advance of £50 by Banks on 18 January 1814.(105) This and several visits Smith made to Banks in 1814 and 1815 were to help "soften the dire aspect of his utter poverty".(106) The publication of Smith's great Geological Map in August 1815, dedicated to Banks; "the most general promoter of science... by his particular encouragement advanced to its present state of perfection", did little to soften Smith's burden of debt, since sales were so soon killed by the rival Greenough version of 1820. Banks subscribed to Smith's *Stratigraphical System of Organised Fossils* in 1817, having opened his magnificent library to aid in its compilation. But this too never proceeded beyond part one, due to a lack of subscribers, despite Banksian urgings.(107)

From 1814 Farey was forced, when writing in the *Philosophical Magazine*, to do so under a cloak of anonymity, in obvious ostracism for his war of words with the gentlemen geologists. The task of lead writer on Geology for Rees's *Cyclopaedia* had already, by 1811, gone to Charles Konig (1774–1851) — "a foreigner holding place of profit here", as Farey wrote.[(108)]

Banks's final attempt to achieve justice for Smith and Farey was on 30 November 1817 when members of the Geological Society met Smith and Farey at a Banks *Conversazione*. Next day Farey drafted a document outlining Smith's discoveries.[(109)] William Henry Fitton (1780–1861), whom Banks had earlier helped to establish in medical practice in Northampton,[(110)] was the Society's mediator. But it was an unsuccessful mediation, and animosity over the issues had reached new heights within the Geological Society by the time of Banks's death.[(111)] In 1822 Farey could refer to the Society as the "*Anti-Smithian Association* in Bedford Street. I have however put their vile attempts *on record* and will yet more effectively do so, please God I live".[(112)]

Smith had been forced to sell his fossil collection to an uninterested British Museum between 1815 & 1818. This final (and present) resting place was due to Banks's crucial intervention,[(113)] since the keeper Charles Konig was a man who in the opinion of Miss Etheldred Benett had "showed an excessive contempt for fossils which so much affronted me on my first acquaintance with him".[(114)] But it failed to solve Smith's financial crisis, and he was in the midst of a ten-week stint in a debtor's prison when Miss Benett wrote thus to Greenough. By 1824 Farey had been reduced to offering himself as a copyist to James Sowerby at a rate of one shilling an hour.[(115)]

Conclusion

It seems paradoxical that the Banks who so perceptively patronised Smith and Farey is so well "patronised" today; whilst they are neglected or denigrated.[(116)] Farey in particular is a polymath who deserves remembrance. In final irony, he was buried in St James Church, Piccadilly, opposite the present home of the Geological Society which had ostracised him. His son, John Farey junior (1791–1851), regarded by some in 1837 as "the best consulting engineer in England",[(117)] was blackballed from the Royal Society in 1847[(118)] as if to show that class divisions remained a critical facet of English history. Engineers were tradesmen. Excluding tradesmen was one of the first "reforms" that Banks had instigated at the Royal Society in 1778.[(119)]

Acknowledgements

I owe grateful thanks to Stuart Band, Julia Bruce, Harold Carter, Roger Flindall, Beryl Hartley, Jean Jones, Oliver and Myra Morgan, Stella Newton and Mick Stanley for kind assistance. Staff at the Sutro Library in San Francisco, the Cornwall and Derbyshire Record Offices and the Derby Local Studies Library provided every assistance. BP Exploration gave financial support without which this contribution would have been impossible.

Notes

(1) K. Garlick and A. Macintyre (eds.), *The Diary of Joseph Farington*, New Haven, 1978, vol. 2, 478.

(2) A.H. Dupree, *Sir Joseph Banks and the Origins of Science Policy*, Minneapolis, 1984.

(3) H.B. Carter, *Sir Joseph Banks 1743–1820*, London, 1988, 26.

(4) S.G. Perceval, '[Banks's] Journal of an excursion to Eastbury and Bristol 1767', *Proceedings of the Bristol Naturalists Society*, n.s.(1899), **9**, 18.

(5) M. Neve and R. Porter, 'Alexander Catcott: glory and geology', *British Journal for the History of Science* (1977), **10**, 37–60.

(6) Carter, op. cit. (3), 49, 53–4.

(7) Ibid., 54.

(8) J.C. Beaglehole, *The Voyage of the Endeavour 1767–1771*, Cambridge, 1955, 518–9.

(9) T.G. Vallance, 'Sydney earth and after: mineralogy of colonial Australia 1788–1900', *Proceedings of the Linnean Society of New South Wales* (1986), **108**, 150.

(10) S. Tomkeieff, 'The basalt lavas of the Giant's Causeway district of Northern Ireland', *Bulletin Volcanologique*, ser. 2 (1940), **6**, 89–143; M. Angelsea and J. Preston, 'A philosophical landscape': Susanna Drury and the Giant's Causeway, *Art History* (1980), **3**, 252–73.

(11) F. Ellenberger, 'Précisions nouvelles sur la découverte des volcans de France', *Histoire et Nature* (1978), **12–13**, 3–42.

(12) R. Pococke, 'A farther account of the Giant's Causeway', *Philosophical Transactions of the Royal Society* (1753), **48**, 230.

(13) K. Taylor, 'Nicolas Desmarest and geology in the 18th century', in *Toward a History of Geology* (ed. C.J. Schneer), Cambridge, Mass., 1969, 346–9.

(14) R.E. Raspe, 'A letter... containing a short account of some basalt hills in Hassia', *Philosophical Transactions of the Royal Society* (1772), **61**, 580–583; A.V. Carozzi, 'Rudolf Erich Raspe and the basalt controversy', *Studies in Romanticism* (1969), **8**, 235–50.

(15) In T. Pennant's [Second] *Tour in Scotland* in 1772 published in 1774, reprinted in J. Pinkerton, *A General Collection of the best and most interesting Voyages and Travels*, London, 1809, vol. 3, 171–569, and R.A. Rauschenberg, 'The journals of Joseph Banks' Voyage up Great Britain's West Coast and to the Orkney Isles, July to October 1772', *Proceedings of the American Philosophical Society* (1973), **117**, 186–226.

(16) R. Porter, *The making of Geology*, Cambridge, 1977, 162, 266.

(17) Anon., 'Sir Joseph Banks', in *Public Characters of 1800–1801*, London, 1801, 380.

(18) British Library, Add. MS. 15510, 22, 34.

(19) Garlick and Macintyre, op. cit. (1), vol. 1, 113.

(20) C. Daubeny, *A Description of ancient and extinct Volcanoes*, London, 1826, 449–51.

(21) J. Pinkerton, *A General Collection of the best and most interesting Voyages and Travels*, London, 1808, vol. 1, 622–3, 730–1.

(22) B. Faujas de Saint Fond, *A journey through England and Scotland to the Hebrides in 1784* (ed. A. Geikie), Glasgow, 1907, vol. 1, 82–3.

(23) Carter, op. cit. (3), 572.

(24) Vallance, op. cit. (9), 151.

(25) T.G. Vallance and D.T. Moore, 'Geological aspects of the voyage of HMS *Investigator* in Australian waters 1801–5', *Bulletin of the British Museum (Natural History), Historical Series* (1982), **10** (1), 5.

(26) H.S. Torrens, 'Under royal patronage: the geological work of John Mawe (1766–1829) and the background for his "Travels in Brazil"', in *O Conhecimento Geologico na America Latina* (eds. M.M. Lopez and S. Figueiroa), Campinas, 1990, 103–113; T.G. Vallance and H.S. Torrens, 'The Anglo-Australian traveller Robert Townson and his map of Hungarian "Petrography"', in *Contributions to the History of Geological mapping* (ed. E. Dudich), Budapest, 1984, 391–8.

(27) Vallance and Moore, op. cit. (25), 2.

(28) John Hawkins to Philip Rashleigh, 14 September [1800], Cornwall County Record Office, DDR 5757/1/109.

(29) *Derby Mercury*, 22 November 1792, 4c.

(30) Anon, 'The family of Hodgkinson, of Overton Hall, in Ashover', *Reliquary* (1872), **12**, 254–5.

(31) H.S. Torrens, 'A Wiltshire pioneer in geology and his legacy — Henry Shorto III (1778–1864)', *Wiltshire Archaeological and Natural History Magazine* (1990), **83**, 176–7.

(32) G.G. Hopkinson, 'Lead mining in 18th century Ashover', *Journal of the Derbyshire Archaeological and Natural History Society* (1952), **72**, 1–21; S. Band, 'Lead mining in Ashover', *Bulletin of the Peak District Mines Historical Society* (1975–6), **8**, 113–5, 129–39.

(33) T.G. Vallance, 'Jupiter Botanicus in the Bush: Robert Brown's Australian field-work 1801–5', *Proceedings of the Linnean Society of New South Wales* (1990), **112**, 51.

(34) J. Milnes, 'Section of the Strata at Gregory's Mine, Ashover, Derbyshire June 26 1801' improved to 1806, Stuart Band collection, Ashover, from an original MS. in Clay Cross Company records.

(35) Ibid.; S. Glover, *The History and Gazetteer of the county of Derby*, Derby, 1833, vol. 2, 52.

(36) California State Library Sutro Branch [hereafter Sutro], MS. Coal 1: 30.

(37) Natural History Museum, London, Dawson Turner copies, vol. 9, 101.

(38) Georgina, a keen mineral collector at nearby Chatsworth, D. King-Hele, *The Letters of Erasmus Darwin*, Cambridge, 1981, 327.

(39) MS. in private hands.

(40) A. Raistrick, *The Hatchett Diary*, Truro, 1967, 7, 11, 65.

(41) King-Hele, op. cit. (38), 174, 267.

(42) J.A. Venn, *Alumni Cantabrigienses*, Cambridge, 1944, part 2, vol. 2, 576.

(43) H.B. Woodward, *The History of the Geological Society of London*, London, 1907, 272.

(44) Raistrick, op. cit. (40), 63–5.

(45) J. Challinor, *A Dictionary of Geology*, Cardiff, 1978, 12–3.

(46) J. Farey, 'Geological observations on the County of Antrim', *Philosophical Magazine* (1812), **39**, 361.

(47) J. Banks, 'Effect of the *Equisetum Palustris* upon Drains', *Communications to the Board of Agriculture* (1800), **2**, 349–50; H.B. Carter, *The Sheep and Wool Correspondence of Sir Joseph Banks*, Sydney, 1979, 305.

(48) H.S. Torrens and T.D. Ford, 'John Farey (1766–1826): an unrecognised polymath, in *General View of the Agriculture and Minerals of Derbyshire* (J. Farey), Matlock, 1989, vol. 1, [reprint, orig. 1811].

(49) Milnes, op. cit. (34).

(50) Copy in the Local Studies Library, Derby, see Band, op. cit. (32), 136–8.

(51) Vallance and Moore, op. cit. (25), 39–43.

(52) Ibid., 3.

(53) S. Band, 'John Allen, miner: on board H.M.S. *Investigator* 1801–4', *Bulletin of the Peak District Mines Historical Society* (1987), **10**, 67–78.

(54) Vallance and Moore, op. cit. (25), 4–5, 9.

(55) J. Phillips, *Memoirs of William Smith*, London, 1844, 39.

(56) J.M. Eyles, 'William Smith, Sir Joseph Banks and the French Geologists', in *From Linnaeus to Darwin* (eds. A. Wheeler and J.H. Price), London, 1985, 37–50.

(57) P. Thompson, *Collections for a topographic and historical account of Boston*, London, 1820, 297–8; W.H. Fitton, *Notes on the progress of Geology in England*, London, 1833, 42–3.

(58) J. Challinor, 'The progress of British geology during the early part of the nineteenth century', *Annals of Science* (1970), **26**, opp. 178.

(59) Oxford University Museum, Smith archive.

(60) Eyles, op. cit. (56), 43.

(61) Anon, 'Agriculture' [report on the Woburn sheep-shearing], *Agricultural Magazine* (1802), **6**, 466.

(62) Eyles, op. cit. (56), 43–4.

(63) Anon, 'Woburn sheep-shearing', *Holly leaves* (1949), Christmas, 12–3.

(64) Phillips, op. cit. (55), 45; L.R. Cox, 'New light on William Smith and his work', *Proceedings of the Yorkshire Geological Society* (1942), **25**, 15, 37, plate 5; Carter, op. cit. (3).

(65) Bedfordshire Record Office, 2114/456.

(66) Cox, op. cit. (64), 38.

(67) Ibid., 18.

(68) H.S. Torrens, 'The transmission of ideas in the use of fossils in stratigraphic analysis to America 1800–1840', *Earth Sciences History* (1990), **9**, 109.

(69) Cox, op. cit. (64), 21.

(70) Ibid., 18.

(71) T.D. Ford, 'The first detailed geological sections across England by John Farey 1806–8', *Mercian Geologist* (1967), **2**, 41–9.

(72) Devizes Museum, Cunnington MS.

(73) Geological Society of London, Minutes of the Society, MS.

(74) Thompson, op. cit. (57), 293–302.

(75) J. Farey, 'On the supposed fresh-water origin of the Gypsum Strata in the environs of Paris', *Philosophical Magazine* (1810), **35**, 259.

(76) Ford, op. cit. (71).

(77) British Geological Survey, 1/1313; Oxford University Museum; Natural History Museum, London and Sheffield Central Library, Oakes deed 1221.

(78) Anon, [note of exhibition of Farey's section]. *Agricultural Magazine* n.s. (1808), **2**, 229.

(79) M.J.S. Rudwick, 'The Foundation of the Geological Society of London', *British Journal for the History of Science* (1963), **1**, 325–55.

(80) Sutro, M 2: 24.

(81) Sutro, Geol. 1: 2a.

(82) Sutro, M 2: 22.

(83) J. Farey, *General View of the Agriculture and Minerals of Derbyshire*, London, 1811, vol. 1, v.

(84) Sinclair to Farey, Sinclair to Banks, 14 August 1807, Sutro, Ag 3: 43/44.

(85) J. Farey, 'Notes and observations on.... Bakewell's *Introduction to Geology*', *Philosophical Magazine* (1814), **43**, 334.

(86) W.R. Dearman and S. Turner, 'Models illustrating John Farey's figures of Stratified Masses', *Proceedings of the Geologists Association* (1983), **94**, 97–104.

(87) Farey, op. cit. (83), vol. 1, 116.

(88) P. Riden, 'Joseph Butler. Coal and Iron Master 1763–1837', *Derbyshire Archaeological Journal* (1984), **104**, 87–95.

(89) J. Farey, 'An account of the Great Derbyshire Denudation', *Philosophical Transactions of the Royal Society* (1811), part 2, 242–56.

(90) Rudwick, op. cit. (79).

(91) Carter, op. cit. (3), 532.

(92) University College, London, Greenough MS.

(93) J. Farey, *General view of the Agriculture and Minerals of Derbyshire*, London, 1813, vol. 2, ix.

(94) Farey to James Sowerby, 20 August 1811, Bristol University Library, Eyles MS.

(95) Sutro, M 2: 25.

(96) T. Thomson, *History of the Royal Society*, London, 1812, 212.

(97) Anon, 'Report of Mr John Farey's paper on the Ashover denudation', *Philosophical Magazine* (1813), **41**, 303–5.

(98) J. Farey, *General view of the Agriculture and Minerals of Derbyshire*, London, 1817, vol. 3, vi–vii.

(99) W. Bainbridge, [Plagiarism of Farey's section by White Watson], *Monthly Magazine* (1818), **45**, 10, 218.

(100) D.P. Miller, Method and the micropolitics of science, in *The Politics and Rhetoric of Scientific Method* (eds. J.A. Schuster and R.R. Yeo), Dordrecht, 1986, 239.

(101) G. De Beer, *The Sciences were never at War*, London, 1960, 191.

(102) Carter, op. cit. (3), 514–7.

(103) Ibid., 537.

(104) J.B. Morrell, 'Professionalisation', in *Companion to the History of Science* (ed. R.C. Olby), London, 1990, 988.

(105) According to Smith's diary (Oxford University Museum).

(106) Phillips, op. cit. (55), 76

(107) Cox, op. cit. (64), 57–8.

(108) Farey to Greenough, 15 June 1813, University College, London, Greenough MS.

(109) J. Farey, 'Mr Smith's geological claims stated', *Philosophical Magazine* (1818), **51**, 173–80.

(110) W.R. Dawson, 'Supplementary letters of Sir Joseph Banks', *Bulletin of the British Museum (Natural History), Historical Series* (1962), **3** (2), 63.

(111) H.S. Torrens, 'The scientific ancestry and historiography of "The Silurian System"', *Journal of the Geological Society of London* (1990), **147**, 659.

(112) Farey to Sowerby, May 1822, Smithsonian Institution, Washington, MS.

(113) J.M. Eyles, 'William Smith: the sale of his geological collection to the British Museum', *Annals of Science* (1967), **23**, 209–10.

(114) Etheldred Benett to G.B. Greenough, 22 June 1819, Greenough MS.

(115) Farey to Sowerby, 26 March 1824, Eyles MS.

(116) R. Laudan, *From Mineralogy to Geology*, Chicago, 1987, 168.

(117) B. Bowers, *Sir Charles Wheatstone*, London, 1975, 111.

(118) Royal Society, London, Archives, Mss. certificates.

(119) J. Barrow, *Sketches of the Royal Society*, London, 1849, 33, 39.

(120) British Library, Add. MS. 15510, 22, 34, through the kindness of Beryl Hartley.

(121) Raistrick, op. cit. (40), 63.

(122) Derby Local Studies Library.

(123) Academy of Natural Sciences, Philadelphia.

(124) Copy from Derby Local Studies Library, MS 9626, through the kindness of Mick Stanley.

(125) Author's collection.

(126) Sutro, M2: 25.

(127) Notes [thus] are from the 1 inch British Geological Survey map Chesterfield Sheet no. 112, reprinted 1971.

D. Knight, 'The application of enlightened philosophy: Banks and the physical sciences', in *Sir Joseph Banks: a global perspective* (eds. R.E.R. Banks and others), Royal Botanic Gardens, Kew, 1994, 77–86.

6. THE APPLICATION OF ENLIGHTENED PHILOSOPHY: BANKS AND THE PHYSICAL SCIENCES

DAVID KNIGHT

University of Durham

On 30 October 1815 Banks wrote to Humphry Davy on hearing about his newly-invented safety lamp for coal miners[1]:

"Many thanks for your kind letter, which has given me unspeakable pleasure. Much as, by the more brilliant discoveries you have made, the reputation of the Royal Society has been exalted in the scientific world, I am of the opinion that the solid and effective reputation of that body will be more advanced among our cotemporaries [sic] of all ranks by your present discovery, than it has been by all the rest. To have come forward when called upon, because no one else could discover means of defending society from a tremendous scourge of humanity, and to have, by the application of enlightened philosophy, found the means of providing a certain precautionary measure effectual to guard mankind for the future against this alarming and increasing evil, cannot fail to recommend the discoverer to much public gratitude, and to place the Royal Society in a more popular point of view than all the abstruse discoveries beyond the understanding of unlearned people. I shall most certainly direct your paper to be read at the very first day of our meeting. We should have been happy to have seen you here; but I am still happier in the recollection of the excellent fruit which was ripened and perfected by the very means of my disappointment, your early return to London."

This magnificent epistle allows us to raise the uncomfortable questions which all those concerned with Banks and the physical sciences must feel. Polarity was a fashionable scientific concept during his reign[2], and pairs of polar opposites may illuminate Banks's position. We may contrast an explorer to a natural philosopher; an administrator to an active researcher; and a utilitarian to a theorist or explainer. In these categories, Banks was unusually polarized by comparison with most eminent men of science (especially those concerned with the physical sciences) even in his own day.

J.J.Thomson, writing about his work on the electron, wrote of William Crookes, a predecessor in the field of cathode rays, that "In his investigations he was like an explorer in an unknown country, examining everything that seemed of interest, rather than a traveller wishing to reach some particular place, and regarding the intervening country as something to be rushed through as quickly as possible".[3]

Instead of recording all sorts of interesting observations, Thomson had a theory to test; and devised apparatus which would do the job quickly and elegantly, going straight to the heart of the matter. Banks was literally an explorer, whose scientific training was based upon the descriptive science of botany, and whose real science had been learned on board *HMS Endeavour* with James Cook. Rigorous simplifying, and testing of theories, had little part in his science; as we can see from the letter to Davy. References to "brilliant discoveries" are offset against "solid and effective reputation", and "abstruse" is the word used to characterize Davy's electrochemical work.

Davy was himself, as with laughing gas, sometimes an explorer; but the work which had exalted the Royal Society in the opinion of the scientific world had been his demonstration that chemical affinity was electrical.(4) This research of 1806 had received a prize from the Institut(5) in Paris: it involved careful experiments to establish what Davy had been sure of, that water was decomposed by an electric current into oxygen and hydrogen only, and that everything else that had been noticed by more exploratory chemists was a side-reaction. To this end, Davy required from the Royal Institution apparatus of silver, gold and agate; building up the idea that fundamental science could only be done by men of genius with splendid equipment.(6)

Priestley had been an explorer among the airs or gases he had isolated;(7) the newer chemistry of Lavoisier(8) and Davy required the capacity to rush through intervening country and reach the objective. Banks was of course pleased at the success of his protegé, but resolute and undistracted aiming at a theoretical goal was not his style. My old physics teacher used to talk of "the less exacting discipline of the more descriptive sciences"; which is perhaps another way of emphasizing different casts of mind within the sciences; and may indicate why those involved in physics and chemistry were not always enthusiastic subjects of Banks's learned empire. Natural history and natural philosophy had distinct approaches to nature, which we may forget when we group them all as science. Rutherford may have said that all science is either physics or stamp collecting; he was probably not the first to feel it. But as well as being a descriptive botanist and explorer, Banks was an administrator:(9) the Royal Society filled his life — his home in Soho Square was open to men of science; and, as in the letter to Davy, invitations to his country houses also went to scientific associates.

Political skill is a great gift. It may well be, and Davy found it so, that eminence in scientific discovery is not the best preparation for high administrative office: wrestling with nature in the laboratory is not quite like dealing with the busy world of men. But we have been accustomed to a world in which an office such as the Presidency of the Royal Society goes to someone eminent for research. It involves promoting institutional rather than individual good, and recognising that one's role may well be encouraging rather than active; which can lead to a rather shattering recognition of being middle-aged, whereas physical science is taken to be a game for the young. Davy was 41 when elected to succeed Banks, with a distinguished career behind him: but Banks had been younger (35) when chosen as PRS, and his importance in science came through his long reign. It is perhaps significant that in the portraits at the Royal Society(10), Banks is shown with the mace, the sign of office; whereas Davy stands proudly beside his lamp, the invention which had promoted him, as he put it, to be a general in the army of science. We expect the office to be the crown of a career made in the laboratory or the field; but Banks was different, and much of his authority depended upon his social position as a landed gentleman — which was not welcome to meritocrats like Davy who later wrote of him: "He was a good-humoured and liberal man, free and various in conversational power, a tolerable botanist, and generally acquainted with natural history. He had not much reading, and no profound information. He was always ready to promote the objects of men of science; but he required to be regarded as a patron, and readily swallowed gross flattery. When he gave anecdotes of his voyages, he was very entertaining and unaffected. A courtier in character, he was a warm friend to a good king [George III]. In his relations to the Royal Society he was too personal, and made his house a circle too like a court".(11) This is no doubt unfair and ungrateful; it is very much the way people in universities talk about their professional administrators.

For Banks, as in the letter, references to the Royal Society came naturally; by 1815, he had been President for more than half his life, and it must have seemed to him as to the unsympathetic that "l'état c'est moi". For him, the interests of science were identical to those of the Society. We are accustomed to seeing scientific discovery as the achievement of a person, or perhaps a team; and the various medals and prizes awarded for science enshrine this idea. We recognize that Rutherford's Cavendish Laboratory was more than the sum of its parts, and Research Rating Exercises make us look hard at today's institutions; but Banks's remarks to Davy are still surprising because the good of the Royal Society is quite so prominent. Davy had been Secretary from 1807 until 1812, but he had since then been abroad for a year and a half, and probably was not thinking of the good of the Royal Society when he undertook his research on the lamp; yet Banks wrote to him as to a team member.

There is no doubt that the solid and effective reputation of the Royal Society, and placing it in a more popular point of view, was one of Banks's overriding concerns; and indeed promoting both science and the public understanding of it remains a major task for his successors in every generation. As President for over forty years (Davy, his successor, was born a few days after Banks's election), Banks undoubtedly brought great dedication to his task and saw the Society through various crises;(12) sometimes with help, from for example Henry Cavendish.(13) He has been heavily and not always fairly criticized by later men of science who were operating in a very different world; but his charm and dedication went with the good administrator's horror of rocking the boat. He did indeed keep the Society afloat and on the course he set. By contrast, active researchers, discoverers, are inveterate boat rockers; keeping the balance, while allowing a little pitching and tossing, is crucial if intellectual institutions are to flourish.

Active researchers in Banks's time were sometimes awkward customers; they might, like Joseph Priestley, be political radicals — and the Royal Society gave Priestley little support in his hour of need(14) after the "Church and King" mob had wrecked his house and laboratory in Birmingham. Or they might be provincials, perhaps also nonconformists, like Dalton who saw nothing for himself in Banks's genteel Royal Society. Its very success as a gentleman's club, and metropolitan society, meant that those included were by no means all active, or even very interested, in science; and it did not by any stretch of the imagination fully represent the British scientific community of the day.(15)

Britain is a very small country, and yet science during the Banksian era was subject to a tyranny of distance:[16] in this case, social distance, coupled with metropolitan[17] disdain for the provincial. Banks was sufficiently secure to be able to cross social boundaries, and to get on with the various gardeners and others to whom he was a patron; but the Royal Society was essentially for officers and gentlemen. The socially-mobile Davy,[18] "indebted for his address to the narrowness of his original circumstances", could not but be aware of this; and no doubt that gives an edge to his remarks about Banks. Science in the nineteenth century became a vehicle for those who would rise above the station and its duties into which they had been born, but they could not play a full part in the Royal Society until it lost its *ancien regime* character.[19]

Banks found it hard also to recognize that science might be carried on outside the Royal Society. His opposition to societies devoted to particular sciences, except for the Linnean Society, is well known.[20] The Animal Chemistry Club[21] which brought together physiologists and chemists in what should have been a promising conjunction was allowed only as a subset of the Royal Society; other groups received a Banksian frown, and Davy for example was bullied into withdrawing from the infant Geological Society. This should not be put down to imperialism, *tout court,* for Banks feared that if men of science did not hang together they would hang separately; but it was a failure to read the signs of the times, and aroused fears of decline. Banks, despite his efforts to be fair, in effect represented only one constituency among men of science in Britain.

Decline is always an emotional business. What is clear is that during Banks's reign the leading scientific power was France. This was true right across the board. Able Britons made interesting discoveries, modified French theories, adopted French notations, and did their best to keep their end up; but it was as citizens of a second-class power. 1815, the year of the Safety Lamp and Banks's letter, was the year of Waterloo; and Sadi Carnot, in his pioneer work on thermodynamics, reflected that Britain had won the war because of her industrial power[22] — notably her steam engines. Visitors to defeated France, and especially scientific ones, nevertheless saw a more modern country than their own, with scientific institutions that might profitably be copied.

Davy's safety lamp was not the only one; in response to the same crisis in the industry, George Stephenson had come up with the Geordie lamp, in which access and egress of air was controlled by thin tubes. He had arrived at this lamp by the traditional method of trial and error; whereas Davy had employed the new technique whereby laboratory investigation of the properties of the explosive gas led to the invention of a device — and because he understood the principle, he could do further research on flames and on heterogeneous catalysis. Davy's lamp and his originality were endorsed by the Royal Society, rather as Newton's position had been in the celebrated conflict with Leibniz; Stephenson was outside Banks's empire, and for the Royal Society it was clearly important that credit should go wholly to one of their Fellows.[23] Stephenson later became a famous man through his steam engines; the dispute over safety lamps, partly carried on by supporters of both sides in Thomas Thomson's respectable *Annals of Philosophy*, reveals the social tensions in Regency science and the narrow base of the Royal Society.

Banks's commitment to the Society he administered was a source of strength as well as weakness; but it does mark him off from some of his more creative contemporaries. One scientific society he did back was the Royal Institution;[24] where his investment paid off with Davy, and his illustrious successors. The R.I. was essentially complementary to the Royal Society. They drew upon the same social groups; but whereas Royal Society meetings involved papers being read, usually by the Secretary, to an all-male auditory, the Royal Institution was founded as a centre for attractive lectures. In accordance with their "unprofessional" character, they were open to ladies: science was to be presented as a part of general culture, perhaps[25] to an audience "composed of the gay and idle, who could be tempted to admit instruction only by the prospect of receiving pleasure": the general public, or at least the wealthier part of it.

Moreover, there was in the basement a laboratory; something the Royal Society had always lacked. Here the celebrated demonstration-experiments could be prepared, and research could go on; sometimes, in Davy's case, in public, before a small audience. We think of the Royal Institution in connection with Davy, Thomas Young and Faraday, as a place where fundamental research was carried on at a level the French had to admire. Faraday in particular is remembered for giving up his "professional" chemical analyses done for a fee,[26] in favour of work in electricity and magnetism where the outcome was quite unclear. But this was not what the founders,[27] whose first formal meeting had been in Banks's house, had in mind. They hoped, and Banks saw it in the safety lamp, for fruit; especially on behalf of the landed interest.

Francis Bacon had believed that experiments of light would in the end lead to experiments of fruit, an idea he more clearly expressed as, knowledge is power. In the seventeenth century, there was little direct pay-off from the new science; but by Banks's time mastering nature was a great project, in which all the sciences would have their role. Bacon's *New Atlantis*, a scientific utopia set on a small island, seemed designed for a metropolis set at the hub of a far-flung empire; and the Royal Society might be developed into Salomon's House, the island's seat of power, authority and knowledge, perhaps with Banks as its Prospero. Bacon's cautious, inductive method was also in fashion (in scientific rhetoric if not practice), because it was feared that speculative science, based on systems rather than method, lay behind the dreams of the French revolutionaries of 1789 and succeeding years. Indeed, revolutionary Paris's supremacy in the sciences seemed to indicate that they might be destabilizing. Bacon, contrariwise, had believed that deep knowledge, based upon cautious generalization, would lead to faith in God, and no doubt in the best of constitutions; and in the 1790s this was what people wanted to hear.

At the Royal Institution, the idea had been that artisans would be admitted to the gallery for the lectures; and indeed it had been hoped to do more for them, and to have exhibitions of new machinery. This came to nothing, but Davy on arrival was put to work to study tanning and agriculture. His researches[28] essentially confirmed the best practice, giving it some scientific rationale; and in lectures he urged the importance of science for the economy. His election to the Royal Society followed, in 1803; and in 1805 he was awarded the Copley Medal of the Society for his chemical researches. The Royal Society's highest award thus came to the young Davy for fairly straightforward researches in applied chemistry, in a clear endorsement of useful knowledge.

A year later, he received the French prize for his electrochemical work; this had no obvious utility about it, but cast light on the nature of chemical affinity.

Our distinction between pure and applied science was not made in Banks's day; there was science and there were arts, and all those involved in science hoped that sooner or later their work would improve the arts, fine or useful. Indeed, the systematic pursuit of knowledge without reference to the good of one's fellows did not then seem an evident good. It might be self-indulgence in mere curiosity; a rather childish impulse carried on into what should be maturity. We should therefore be careful about seeing Banks's concern with utility as extreme; but in his letter to Davy the emphasis upon usefulness is as striking as the references to the Royal Society and its public image. Defending society from a scourge was the proper business of natural philosophy.

What distinguished Davy's work from that of Stephenson and many others who were involved in the Industrial Revolution was that it was based in the laboratory, and followed upon the discovery of an underlying principle. It was not just rule of thumb, or inspired analogical thinking. The chemist had to be a practical man, and Banks (welcoming a new instrument in 1816) deplored the way chemists wanted to be gentlemen;[(29)] but effective chemical thought had to be done with the head as well as the hands. Swift's *Voyage to Laputa* had satirized Newton's Royal Society as a collection of learned fools; a sensible man like Banks did not want to preside over such an outfit. Enlightened usefulness was on the other hand a very proper objective for the Royal Society, and prominent among the hopes of its founders in the seventeenth century.

To Sir William Huggins, President a hundred years after Banks at the beginning of our century, it seemed that: "The supreme value of research in pure science for the success and progress of the national industries of a country can no longer be regarded as a question open to debate, since the principle has not only been accepted in theory, but put in practice on a large scale, at a great original cost, in a neighbouring country, with the most complete success.[(30)] What Huggins saw in Germany, Banks could not have seen in France a hundred years before. British seamen apparently preferred captured French warships because they were better designed than British ones;[(31)] under Lavoisier French gunpowder had been superior to British; and French bridges were more elegant. But in general, in Banks's day, British industry and British agriculture were well ahead of French, despite the Parisian lead in the sciences.

Paley's *Natural Theology* was first published in 1802, and the appeal of its vision of the world as a watch[(32)] seems to owe something to the newly developed marine chronometer[(33)] so vital to the explorer. But while Banks no doubt believed in a Creator, the idea that a major role of science should be finding out about God[(34)] through his works seems to have been alien to him. It could bring seriousness to the otherwise apparently frivolous inquiry into the workings of nature; and many in Banks's generation were affected by the Evangelical Revival, initiated by the Wesleys. Nevertheless, faced with childlessness and the gout, Banks apparently found an undergoing stomach to bear up against what might ensue not in the consolations of religion, but in hard work and domesticity. It was straightforward usefulness that he sought from the science over which he presided.

The years around 1800 were a time of poor harvests and of dearth; in which the Faraday family for example found themselves very near the breadline, and the eminent Quaker chemist William Allen distributed relief to the poor. Banks's hope was that work such as Davy's on fertilisers and insecticides would lead to agricultural productivity; and this was very close to his own interests in naturalizing animals and plants in regions new to them. Benevolence and scientific usefulness[35] went readily together, as the association of Sir Thomas Barnard and others like him with the Royal Institution shows. Hunger was undoubtedly an evil and also a threat to society; and it seemed that science, especially in the form of organized common sense in which it was so important to Banks (and right through the nineteenth century), might abolish famine and much disease. Davy indeed wrote[36] that "science is nothing more than the refinement of common sense making use of facts already known to acquire new facts": although for him this was not the whole story.

Davy's safety lamp went beyond organized common sense; and it was also far more dramatic than steady improvements in crop yields could ever be. The chronometer, invented by a clockmaker and a not a man of science, was a mechanical device that would save lives by indicating to sailors where they were; but here was something depending upon the latest chemical and physical science, which would directly and obviously save lives. Metropolitan science at last had something to boast about. Banks could feel unspeakable pleasure at this vindication of long-standing hopes.

The picture we get from the letter is of the natural philosopher not closeted in an ivory tower, but ready to come forward when called upon to defend society from a tremendous scourge of humanity. The enlightened chemist is a new St. George, bodly tackling an alarming and increasing evil — the experiments were dangerous, and in the Royal Institution's laboratory, safety precautions usually only seem to have been taken when there had in the past been an explosion. Because he is applying enlightened philosophy, he can (as St. George could not) promise that he has a "certain precautionary measure effectual to guard mankind for the future"; he knows the principles behind the device. Banks's letter notes that the lamp "cannot fail to recommend the discoverer to much public gratitude": the Royal Society's came in the form of the Rumford Medal, which is awarded biennially for the most important discoveries made in heat and light. This seems fair enough, especially because Rumford had distinguished himself in the invention of economical stoves, and in the founding of the Royal Institution; and had intended to encourage[37]: "such practical improvements in the management of heat and light as tend directly and powerfully to increase the enjoyments and comforts of life, especially in the lower and more numerous classes of society".

The public, or rather the powers that be, showed their gratitude in the form of a baronetcy; raising Davy to the same rank as Banks, and a higher recognition than Newton had received — scientific peerages were still a long way off. Nevertheless, to Davy as to Banks, it was an empty honour to a childless man; and as a reward for saving lives in the year of Waterloo, it contrasted with the more generous honours poured upon the soldiers and sailors who had destroyed so many of their fellow-men.[38] Nevertheless, useful science undoubtedly did win gratitude; and in promoting it, Banks did not differ from Rumford (who had received the first of his own medals), or Davy, or others involved in the physical sciences.

This must surely be the key to understanding Banks's view of the physical sciences. One who had sailed with Cook could not doubt the value of astronomy, although he opposed the setting up of a society dedicated to it; and he promoted the Board of Longitude, dedicated to the practical use of astronomy in navigation. In Banks's day, as we can see from the arrangement of the *Encyclopedia Metropolitana*, planned by Coleridge about the time Davy was inventing his lamp, the "pure sciences" were those without an empirical component: — pure mathematics, and logic. All the rest were "mixed sciences", chemistry and mechanics along with manufactures. Science was distinguished from mere practice, based upon rule of thumb and associated with unreasoned resistance to change;[39] and was expected to be useful, sooner or later. Nature was there to be mastered; but those like Banks who had little patience with abstruse investigations would miss some of the pleasures and satisfactions of the physical sciences. Unlike Davy, he would not readily have seen the chemist as "animated by a spark of the divine mind", following a pursuit which exalted the understanding but did not depress the imagination.[40]

Banks thus emerges as an explorer, an administrator of science, and a promoter of useful knowledge. His emphases were a bit different from those of contemporaries devoted to the physical sciences, but not wildly different: Priestley invented soda-water, W.H. Wollaston manufactured platinum apparatus, Davy invented his lamp. It was a younger generation, including those nowadays called the Cambridge Network,[41] who began to move away from the kind of utilitarian emphasis natural to Banks. If science was to be a part of a liberal education, then it must be the theoretical parts which take precedence. Gradually too the lesson of the safety lamp was assimilated, and the model of technology as applied science, so boosted by Huggins, accepted as a reasonable picture of what normally happens. It did fit the new electrical and dye industries, but by no means all the others. Our view of Banks in relation to the physical sciences will depend upon what we require of a President of the Royal Society: he had little empathy with discoverers in these fields, but was prepared to give them a fair measure of the Society's support, and if their work could benefit humanity then he would be enthusiastic about it. Enlightened philosophy was his delight; and that is no bad thing for a man in his position.

Notes

(1) J. Davy, *Fragmentary Remains, Literary and Scientific, of Sir Humphry Davy*, London, 1856, 208.

(2) A. Cunningham and N. Jardine (eds.), *Romanticism and the Sciences*, Cambridge, 1990.

(3) J.J. Thomson, *Recollections and Reflections*, London, 1936, 379.

(4) D.M. Knight, *Humphry Davy: Science and Power*, Oxford, 1992, chap. 5; J.Z. Fullmer, *Sir Humphry Davy's Published Works*, Cambridge, Mass., 1969.

(5) M.P. Crosland, *Science under Control: the French Academy of Sciences, 1795–1914*, Cambridge, 1992, 23ff.

(6) J. Golinski, *Science as Public Culture: Chemistry and Enlightenment in Britain, 1760–1820*, Cambridge, 1992.

(7) On the Chemical Revolution, see W.H. Brock, *Fontana History of Chemistry*, London, 1992.

(8) A. Donovan, *Antoine Lavoisier*, Oxford, 1992.

(9) H. Lyons, *The Royal Society, 1660–1940: a History of its Administration under its Charters*, Cambridge, 1944, chap. 6, esp. 198ff.

(10) N.H. Robinson and E.G. Forbes, *The Royal Society Catalogue of Portraits*, London, 1980, 18, 82.

(11) J. Davy, *Memoirs of the Life of Sir Humphry Davy*, London, 1836, vol. 2, 126.

(12) H.B. Carter, *Sir Joseph Banks*, London, 1988, chap. 9; this biography is invaluable.

(13) R. McCormmach, 'Henry Cavendish on the proper method of rectifying abuses', in *Beyond History of Science* (ed. E. Garber), Bethlehem, Pa., 1990, 35ff.

(14) On Priestley, see the historical papers in Royal Society of Chemistry, *Oxygen and the Conversion of Future Feedstocks*, London, 1983, (special publication, 48).

(15) J. Morrell and A. Thackray, *Gentlemen of Science: the early years of the B.A.A.S.*, Oxford, 1981, chaps. 1–3.

(16) D.M. Knight, 'Tyrannies of distance in British Science', in *International Science and National Scientific Identity*, (eds. R.W. Home and S.G. Kohlstedt), Dordrecht, 1991, 39–53.

(17) I. Inkster and J. Morrell (eds.), *Metropolis and Province*, London, 1983, esp. chap. 1.

(18) Obituary of Davy, *Annual Register* (1829), **71**, 505.

(19) M.B. Hall, *All Scientists Now: the Royal Society in the Nineteenth Century*, Cambridge, 1984.

(20) For a view a hundred years on from Banks, see W. Huggins, *The Royal Society*, London, 1906, chap. 2.

(21) See the papers by N.G. Coley in *Notes and Records of the Royal Society*,(1967), **22**, 173–85, and *Ambix* (1988), **35**, 155–68.

(22) S. Carnot, *Reflexions sur la puissance motrice du feu* (ed. R. Fox), Paris, 1978 [1824], 62.

(23) S. Smiles, *The Lives of George and Robert Stephenson*, reprint, London, 1975, [1874], chap. 6.

(24) There are some MSS relating to Banks's involvement in the early days at the Royal Institution, and I am very grateful to the Librarian for supplying me with xeroxes of them.

(25) Obituary of Davy, *Annual Register* (1829), **71**, 507.

(26) J. Tyndall, 'Faraday as a Discoverer', [1868], reprinted in Royal Institution Library of Science, *Physical Sciences*, Amsterdam, 1970, vol. 2, 116ff.

(27) M. Berman, *Social Change and Scientific Organization: the Royal Institution, 1799-1844*, London, 1978, chap. 1.

(28) D. Knight, *Ideas in Chemistry: a History of the Science*, London and New Brunswick, NJ, 1992, chap. 8.

(29) Carter, op. cit. (12), 517.

(30) Huggins, op. cit. (20), 21.

(31) Society of Arts, *Lectures on the Results of the Great Exhibition of 1851*, London, 1852, 547.

(32) J.H. Brooke, *Science and Religion: some historical perspectives*, Cambridge, 1991, chap. 6.

(33) D. Howse, *Nevil Maskelyne: the seaman's astronomer*, Cambridge, 1989, 122–7.

(34) D. Knight, *The Age of Science: the scientific world view in the nineteenth century*, Oxford, 1986, chap. 3.

(35) Berman, op. cit. (27), chap. 1.

(36) Knight, op. cit. (4), 44.

(37) Lyons, op. cit. (9), 219.

(38) Davy, op. cit. (1), 210.

(39) R.F. Bud and G.K. Roberts, *Science versus Practice: Chemistry in Victorian Britain*. Manchester, 1984.

(40) H. Davy, *Collected Works,* London, 1839–40, vol. 9, 361; part of a dialogue concerned with "chemical philosophy".

(41) S.F. Cannon, *Science in Culture*. New York, 1978, chap. 2.

R. Joppien, 'Sir Joseph Banks and the world of art in Great Britain', in *Sir Joseph Banks: a global perspective* (eds. R.E.R. Banks and others), Royal Botanic Gardens, Kew, 1994, 87–103.

7. SIR JOSEPH BANKS AND THE WORLD OF ART IN GREAT BRITAIN

RÜDIGER JOPPIEN

Museum für Kunst und Gewerbe, Hamburg

Until the age of thirty, Banks was a man of action, a traveller to Newfoundland, to the South Seas and to Iceland, but his later years were comparatively uneventful. He began as a scholar-traveller, but most of his life was spent instigating and receiving the scholarship of others.

In his services to the science and commerce of his country, Banks was certainly as great a figure as say Boulton, Wedgwood, even Cook; however, his achievements were not charismatic, being mostly administrative, and they have not been held in awe by posterity. What is still visible today, if only to the initiated, is Banks's herbarium in the Natural History Museum, and his books in the British Library, which bear his bookplate. Compared with his contemporary, Sir Charles Towneley, represented in a painting by John Zoffany sitting amongst his collection which now belongs to the Department of Greek and Roman Antiquities, Banks's achievements do not seem spectacular. Two hundred years ago, his house at 32 Soho Square, served as one of the world centres of scholarship, possessing a research library with a truly international standing, and being the repository of the most varied collections of plants, animals, fossils, minerals, artefacts, books, and all kinds of pictorial documents.

The role of Banks as a scholar and a promoter of scholarship is gradually becoming better known. This is due to Harold Carter's monumental biography of Banks which has given new insights into the web of activities, science, commerce, agriculture, etc., in which Banks was involved. One of the still little known aspects of his accomplishments, is that of his role as a patron and a collector of works of art.

One only has to look at Banks's life to realise that he had contacts with many artists of the day, and that these ranged from the most elevated to quite modest ones. We have a number of portraits of Banks, which originated not necessarily from his own vanity, but naturally from his role as a public man. Quite a few are in pencil, like an early one by Thomas Hearne, executed at a time when Hearne was still a pupil working with the engraver William Woollett. Others are more representational, in oil, and Banks may well hold the record of having been portrayed by three consecutive presidents of the Royal Academy: Reynolds, West and Lawrence.

If one looks at Banks's portraits by Benjamin West (Fig. 1) and Joshua Reynolds, it is obvious that these pictures convey the message of an active man. Pointing to his Maori garment, woven from the flax plant, Banks shows his practical strain, for he favoured the introduction of the New Zealand plant into England. In Reynolds's half-length picture with the globe on the table, Banks is

FIG. 1. Benjamin West. Joseph Banks in a Maori Cloak, 1771. Oil on canvas. Reproduced by permission of Lincolnshire County Council: Usher Gallery, Lincoln.

epitomised as the restless traveller. John Russell's crayon portrait of Banks of 1788 makes one aware of Banks's interest in astronomy and his support in the drawing of the lunar map, which Russell had begun in 1785.

For a portrait by Thomas Phillips of 1808, commissioned by Banks's friend, the astronomer and mathematician José Mendoza y Rios, Banks himself suggested that he should hold the Bakerian lecture by Humphry Davy of 1808 in his hand. The portraits by Thomas Phillips of 1810 and 1814 point to more everyday utilitarian considerations such as the curing of the wheat disease, on which Banks wrote a pamphlet, shown on the table, or the drainage of the fens in his native Lincolnshire, a plan of which Banks holds in his left hand. Such was his idea of useful communication, that everything that he pursued had to have a practical end.

What I would call the utilitarian strain is also apparent in other pictures which Banks commissioned, such as the two portraits of Captain Cook and Captain Charles Clerke, which were executed by Nathaniel Dance. While Cook is reading the chart of the Pacific Ocean which he himself had so substantially enlarged, Captain Clerke and the Maori chief behind him (the latter derived from one of Parkinson's drawings) seem to discuss, again, the advantage of New Zealand flax.

Four years earlier (1772) Banks had commissioned the animal painter George Stubbs to do portraits of a kangaroo and a dingo, of which Banks had brought home skulls and skins. In having them painted, Banks followed a precedent set the year before by the naturalist William Hunter, who had commissioned Stubbs to paint a picture of the Duke of Richmond's first bull moose, recently sent over from Canada. In his paper read to the Royal Society, Hunter stressed the point of having the moose painted by Stubbs, with the words "Good paintings of animals give much clearer ideas than descriptions".

The value of visual representations were of course also appreciated for the depiction of people from foreign lands. The most noteworthy visitor to England in the second half of the 18th century was the Tahitian Omai, whom Banks, after Capt. Tobias Furneaux had brought him back from Capt. Cook's second voyage in 1775, placed under his personal protection — until Cook took him home to his native land. During his stay in England Omai was courted — and depicted, monumentally by Reynolds (on Banks's commission, we may ask?) and more modestly by Nathaniel Dance (Fig. 2). From the latter's drawing Francesco Bartolozzi produced a celebrated engraving, published in 1775.

Throughout his life Banks was attached to the keeping and welfare of foreign visitors to Britain, and often took the chance of having their pictures taken. Two portraits of the Inuit Attuiock and Caubvick, who came to England in 1773, were drawn by Nathaniel Dance — and twenty years later copied by Christopher William Hunneman (1730–93) as breast portraits. These copies Banks sent to his eminent scientific friend, the anthropologist Johann Friedrich Blumenbach from Göttingen, and they are now in the Völkerkundliche Sammlung Göttingen. Their inscription on the back reads (for Attuiock): "An Esquimaux man who was brought over from Cape Charles on the Coast of Labrador by Capt. Cartwright a. 1773. He was a priest in his country, his name Ettuiack. The original drawing in the possession of Lady Banks was made by Nathaniel Dance 1773. This copy by Mr Hunneman 1792".

In spring 1797 Banks also gave a picture, or rather the copy of a picture, of

FIG. 2. Nathaniel Dance. Portrait of Omai, 1775. Pencil. Reproduced by permission of the National Archives of Canada, Ottawa

three natives of the Palau Islands to Blumenbach. This was painted in Macao, in 1791, and came subsequently into Banks's possession as well as a portrait of the Inuit woman Mikok and her son Tootac. The picture, which is still in Göttingen, has the following inscription (translated from the German): "Mycock, an Esquimaux woman with her son, who were taken from Labrador to London by Lieut. Francis Lucas in 1769 and who were painted by John Russell for Sir Joseph Banks, who gave it to me as a present".

The extent of Banks's patronage and employment of artists and the scope of his pictorial collections have not yet been fully examined. As yet, only some of the contours are visible. With Banks the element of employment of artists and the engagement in collecting cannot be readily separated, for one arose out of the other. It is obvious that the bulk of his commissions were to do with botany, especially when we think of his patronage of Dionysius Ehret in 1768 or the employment of Sydney Parkinson during Cook's first voyage. But from the beginning of setting up his "working" collection, Banks was interested in a wide range of pictorial documents. Among Banks's collection in the British Museum (Bloomsbury) there is a small group of drawings in bodycolour on vellum by Sydney Parkinson, showing delightful studies of insects; they are signed and dated 1768. Other drawings from the same collection, mostly of birds, show that Parkinson's commitment to Banks had begun a year earlier, probably right after the latter's return from Newfoundland, when many dried and stuffed animals had to be painted. It was Parkinson who was Banks's greatest collaborator on his voyage to the South Seas. In the course of the voyage Parkinson, definitely guided by Banks, turned into the predominantly botanical draughtsman whom we know. However, there are also many brilliant drawings by him of fish, birds and artefacts.

There was a period when Banks took great delight in artificial curiosities and set his artists the task of drawing them. In terms of time, this goes up to at least 1774, when John Cleveley, an artist strongly attached to Banks since 1772, made drawings of those artefacts which Captain Furneaux had brought home from Cook's second voyage.

After the death of Parkinson and Alexander Buchan, who was the second artist on the first voyage, John Cleveley and James and John Frederick Miller represented the next generation of Banksian draughtsmen. Their names are connected with a great deal of copying and elaborating of earlier South Sea plant drawings, and also with redrawing some of the images from Captain Constantine Phipps's voyage to Spitzbergen in 1773. Their most valuable contributions however, were their drawings from Banks's voyage to the Hebrides and Iceland in 1772.

Banks was most active as a collector. An impressive survey of his major collection items in the field of natural history drawings was published during Banks's lifetime by his librarian Jonas Dryander in his 5-volume edition of *Catalogus Bibliothecae Historico Naturalis Josephi Banks* (1796–1800).

It is well known that Banks's collection included Charles Plumier's watercolour and pen-and-ink drawings of plants, Jean Baptiste Aublet's 610 original drawings of plants from Guiana and William Bartram's botanical and zoological drawings of the southern states of the United States; furthermore a large body of unauthorised drawings mainly from India and China, by native artists; and finally several collections which Banks acquired through his connections to the voyages and travels of his time. Banks owned the largest and possibly only collection there was of drawings from Cook's three voyages, not only Parkinson's, Buchan's and Spöring's drawings from the first voyage, which he himself instigated, but also four volumes of plants and animals by George Forster from the second, as well as animal drawings by Capt. Charles Clerke, William Ellis and John Webber from the third voyage. In fact, it is possible to say that 95% of the natural history drawings taken during Cook's three voyages were held by Banks.

Of particular interest is Banks's miscellaneous collection of animal drawings (now in the British Museum, Dept. of Prints and Drawings), of which his second librarian, Jonas Dryander, prepared a MS. catalogue in about 1786, held in the Natural History Museum. It was arranged according to the Linnaean systema sexualis, which proved a useful method of classification for any drawing that came Banks's way.

FIG. 3. John William Lewin. A Wombat, N.S.W., 1801. Watercolour. British Museum, Dept. of Prints and Drawings (1914-5-20-283). Reproduced by permission of the Trustees of the British Museum.

Contained in this collection are drawings, such as Aert Schouman's watercolour of a tropical bird, which Banks might have acquired during his travels through Holland in 1773, and John William Lewin's drawing of a wombat (Fig. 3), signed and dated "N.S.W. 1801". From Lieutenant Colonel William Paterson's letter to Banks of 20 August 1801, one learns that he and Lewin had undertaken a journey towards the Hunter River, and that he had sent a drawing of a wombat to Sir Everard Home, Banks's intimate friend and physician; this animal, Paterson says, they had captured and kept alive for a number of days. I tend to believe that this is the one drawn by Lewin.

What do we make of George Stubbs's drawing of Lemurs (Fig. 4), which we know was made — together with two other ones — in 1773, when Banks and Stubbs went to visit Marmaduke Tunstall's menagerie at Leicester Square, and what are we to think of Nathaniel Dance's drawing of a shrimp? I cannot believe that such drawings were the result of proper commissions. Equally Sawrey Gilpin's free and spirited pencil drawing of squirrels may have been an incidental acquisition, while George Garrard's gouache of a stag from the Princes Island — the animal had been sent as a present to Lord Orford in 1797 and also painted by Garrard on canvas — was probably ordered and paid for, as Banks's inscription suggests. Trifling as these examples may seem, I find it interesting that Banks's attention was drawn towards some of the most noteworthy English animal painters of the age: Garrard, Gilpin, Stubbs. A drawing in gouache of a cheetah by Peter Paillou, probably from the late 1760s, is annotated in Banks's hand: "copied from a painting by Stubbs, taken from a living animal in Ld. Piggot's possession" (painted 1765). Thus one wonders whether already at this early stage Banks had become acquainted with the work of Stubbs.

We shall refer back to Banks's role as a collector and patron, after having cast a glance at some of his other activities: his role as a publisher of travel accounts and natural history books, and as an adviser on voyages of exploration. This story, which can only be touched upon here, begins with Banks's own *Florilegium*, which, although about 750 copper plates had been prepared, he abandoned after the death of Daniel Solander in 1782. Just at this time when his own publication would have needed a big push to complete, Banks supervised the publication of Cook's third voyage (1784), with its 61 engraved plates. In terms of organisation and finance the work soon came to be regarded as the model for further publications, and indeed Banks was active in this field up to the last decade of his life.

Several publications, such as *Hortus Kewensis* (3 vols, 1789), *Delineations of Plants cultivated in the Royal Garden at Kew* (1796), *Icones Plantarum quas in Japonia collegit et delineavit Engelbertus Kaempfer* (1791), or William Roxburgh's *Plants of the coast of Coromandel* (3 vols, 1795), ensued from his activities.

Among Banks's greatest achievements was his role as an instigator and adviser upon voyages and travels. Though this has always been acknowledged — so that in 1794 a contemporary correspondent addressed Banks as "the father of research, the laborious advocate of enquiry and the friend of the adventurous traveller" — the full extent of his preoccupations with the voyages and travels of his age have only recently been demonstrated by Harold Carter in his Banks biography. Among the expeditions which Banks either instigated or advised upon were those from the Admiralty to the South Seas and Australia, to the

FIG. 4. George Stubbs. Lemurs, 1773. Pencil. British Museum, Dept. of Prints and Drawings (1914-5-20-304). Reproduced by permission of the Trustees of the British Museum.

North-West Coast of America, or to the Arctic. But Banks also sent travellers to Africa, and he was involved in all three embassies to the Emperor of China between 1787 and 1816. Two of the delegations had artists to accompany them, Julius Caesar Ibbetson on Cathcart's and William Alexander on Lord Macartney's mission.

Julius Caesar Ibbetson's role as draughtsman to Earl Cathcart's embassy to China in 1787/88 was short lived and only went as far as Java. When Cathcart died off Anger Point, Java in June 1788, the expedition was abandoned. Ibbetson thus only had the chance to cover some South African scenery around the Cape, and to document the burial of Cathcart, a drawing, now in the Victoria & Albert Museum, from which subsequently he developed a large painting in oil on canvas, which was exhibited at the Royal Academy in 1789 (No. 188).

When a second embassy was fitted out under the leadership of Lord Macartney (1792–94), Ibbetson was approached again and suggested his student William Alexander to go instead.

Alexander's records were singularly rich. Not since the middle of the 17th century had so many and diverse images about China, its scenery and people, reached the West, and Alexander's work proved a particular asset of the official publication. When the latter was planned, it was the publication of the account of Cook's third voyage that was taken as a model. Again, Banks gave his organisational support, supervising the general shape and finances of the publication.

Banks played an important role in the preparations of Capt. Matthew Flinders's expedition in the *Investigator* to Australia (1801). On the advice of Banks, this expedition included a promising young botanist, Robert Brown, who later became Banks's third librarian, a natural history draughtsman, Ferdinand Bauer, and a landscape artist, William Westall.

Thirty years after Sydney Parkinson, Ferdinand Bauer became the second professional flower painter to explore the natural treasures of Australia. Banks knew from experience that the Australian flora was demanding beyond measure, and a working coalition between Robert Brown and Ferdinand Bauer was set up to render the botanical survey more effective. Banks was not wrong in his expectation, for Bauer returned from Australia with more than 2000 sketches, all of which, when properly finished, gave hope of being, in Banks's words, "beautiful to the eye and interesting to science".

As a landscape and figure draughtsman, Banks had wished for William Daniell, who would have brought great artistic experience to the task from his travels to India. But, when Daniell withdrew at the last minute, Banks at least secured the 20-year-old art novice William Westall for the job.

Westall had been brought up with notions of the picturesque, but found himself sadly disappointed, as he said in a letter to Banks, with the barren Australian coast, which afforded few places of artistic interest. Like Webber, who had worked under the personal supervision of Cook, Westall was placed under the surveillance of Flinders, accompanying him on surveying excursions. Thus among his landscapes we find a great number of panoramic views that were taken from the top of mountains. At a later stage of the voyage natives were encountered on several occasions, at which Westall took portrait-sketches. To these he added a number of drawings of Aboriginal rock paintings.

When Westall returned to England in 1805 his drawings were duly transferred to the Admiralty to await possible publication. Yet again we find Banks involved in the supervision of a published account, which finally appeared in July 1814. For about seven years, from 1805 to 1812, Banks had dealt with Westall's artistic records from the voyage, canvassed on his behalf and reinstated Westall into his old contract with the Admiralty. The result was a considerable number of watercolour drawings and ten paintings, which to the present day adorn the rooms of Admiralty House, London, or are hung on indefinite loan in the National Maritime Museum, Greenwich. Taking Westall's interest in hand, Banks reveals himself as a promoter of the arts, with reliability and responsibility to the cause that was entrusted to him.

Because of the nature of his engagement Banks dealt with all kinds of artists: he mingled with members of the Royal Academy at the same time as superintending the works of the modest engraver. The number of engravers whom Banks employed throughout his life must amount to about 80 according to my estimation, and one would therefore assume that, next to such tycoons of the print business as Boydell, Bowyer and Macklin, Banks was one of the great promoters of prints of his period.

Banks's working knowledge of the print media of his time can be best epitomised by the devising of the vignette for the title-page of the *Philosophical Transactions*, as it was introduced in 1778.

On 24 March 1778 the Council Minutes of the Royal Society read: "Resolved that a drawing be made of the Medal in order to be cut in wood, for an ornament in the title page of the transactions, & that Mr Banks be requested to oversee the execution. Ordered that the thanks of the Council be returned to Mr. Carlberg for his elegant drawing of the Medal which was produced by Mr Banks".

Mr Carlberg has ever since remained an obscure artist — his relationship with Banks is otherwise not documented. Better known is J. Caldwall, the engraver, who to us is well known from his work for the atlas of Cook's third voyage.

What perhaps contributed to Banks's experience of printmaking, was his work in the various societies to which he belonged. James Basire, who worked on the Cook plates, was nominated engraver to the Royal Society in 1770, and Mr Hixon, the printer, equally involved in the Cook atlas, worked for the Society at least temporarily as a plate printer. Basire also had various connections with the Society of Antiquaries and the Society of Dilettanti as an engraver for more than thirty years.

Through the Society of Antiquaries Banks would have got to know the draughtsmen and engravers Francesco Bartolozzi and Giovanni Battista Cipriani, who were made honorary members in 1767; they both became founding members of the Royal Academy in 1768 and contributed much to Banksian publications.

Banks undertook a very different commission when in spring 1789 he asked the historical painter William Hamilton for an allegorical painting to glorify the recovery of the King from illness. It was executed as a transparency and placed in front of the windows on the facade of 32 Soho Square. The subject was afterwards engraved by F. Bartolozzi and published by C. Ansell in June 1790, duly dedicated to "Sir Joseph Banks Bart. President of the Royal Society". (Fig. 5).

Though Banks was not, I believe, much attracted by the fine arts and forms of historical or religious painting, landscape or still-life, he had many contacts with connoisseurs and collectors. On 6 February 1774, on the recommendation of James "Athenian" Stuart, Banks was elected a member of the Dilettanti Society,

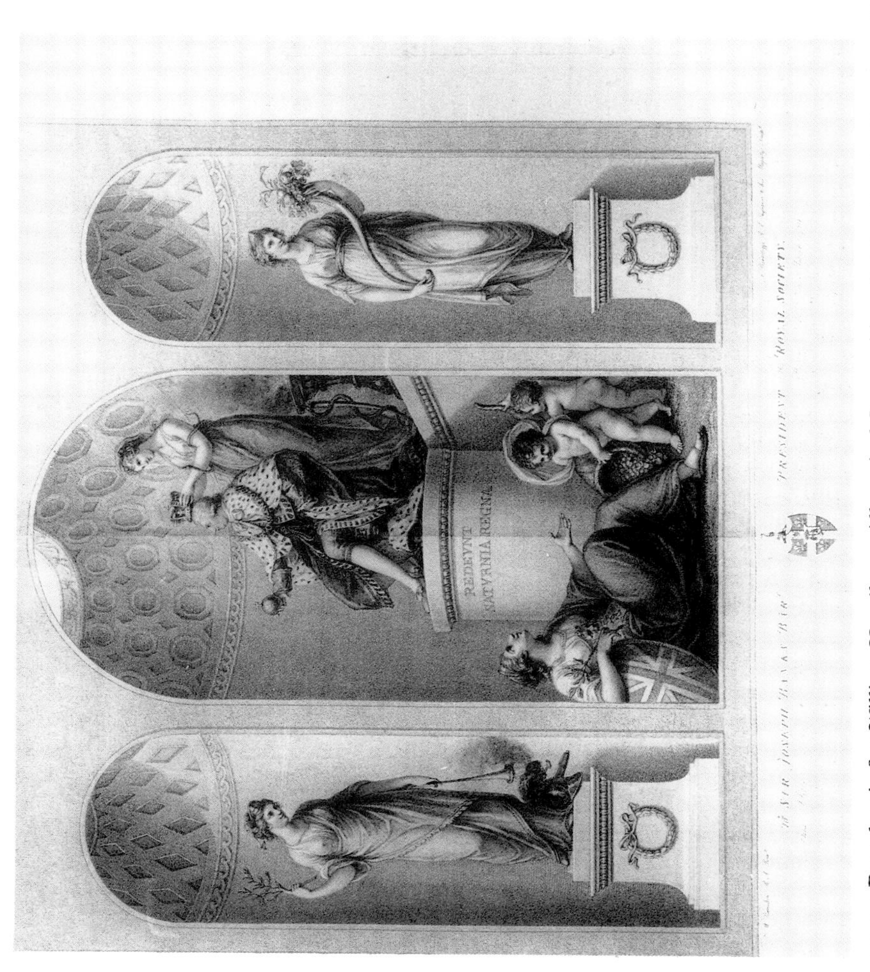

FIG. 5. Francesco Bartolozzi after William Hamilton. Allegorical Composition, Celebrating the Recovery of King George III, with a dedication line to Sir Joseph Banks, President of the Royal Society, 1789. Engraving. Reproduced by permission of Collection David Alexander, York, England and the Paul Mellon Centre for Studies in British Art.

just before his friend Charles Greville, the nephew of Sir William Hamilton. There were a number of other illustrious members, with whom Banks mixed and certainly found agreeable, such as the patron and connoisseur, the Duke of Richmond, Sir Joshua Reynolds, the Earl of Sandwich, Constantine Phipps, Lord Mulgrave, Sir Watkin Williams-Wynn (the Welsh patron of Paul Sandby), Sir William Hamilton, Nathaniel Dance, David Garrick, George Colman, Richard Payne Knight, Charles Towneley, and many others who belonged to the cultural establishment of their time. (Fig. 6).

The Society of Dilettanti, and particularly its member Charles Greville, were instrumental in deepening Banks's interest in the antique, and there is every reason to believe that classical antiquity interested Banks not so much on an aesthetic and cultural but on an archaeological level. Banks was a friend of Sir William Hamilton and knew his collection. He was not less familiar with Richard Payne Knight, who in 1787 published his *Account of the Worship of Priapus*, which contained Sir William Hamilton's letter to Banks as President of the Royal Society on the cult of Priapus (1781) as it was discovered at Isernia in Abbruzzo.

Together with Charles Towneley, Banks sat as a Trustee of the British Museum ex officio; and Antony Griffiths, the Keeper of Prints and Drawings of the British Museum, kindly informed me that, judging from "the evidence of the attendance at the meetings ... one can certainly say that Banks was one of those most consistently present" (21.8.91). Thus, without going into any further detail here, one may conclude that from Stuart and Revett, Richard Chandler, Sir William Hamilton up to Charles Towneley and Richard Payne Knight, Banks was well aware of and in friendly intercourse with all late 18th-century and early 19th-century British champions of the Greek revival.

On a number of occasions Banks met people like Wedgwood, Thomas Bentley and J. Flaxman. When Banks moved into 32 Soho Square in 1778, the "council chamber" was furnished "with a magnificent yellow marble chimney-piece set with a green jasperware plaque of the 'Offering to Flora'. The design dates from 1778 and was based on a cast from Arbury Hall, probably modelled by Hackwood. John Flaxman adapted the plaque to the mantelpiece, making the necessary adjustments in size".

Banks in turn had the opportunity of hosting Wedgwood at a special occasion at his house in early May 1790, when Wedgwood came down to London in order to display the replica of the famous Portland vase among a small gathering of friends. After Flaxman's departure for Rome in 1787, Wedgwood had found another able designer in the person of Henry Webber, who modelled the Portland vase between 1787 and 1789. That Henry Webber was the brother of John Webber, Cook's artist on his third voyage, is but a coincidence, but may just underline the intensive relationship of Banks with the London art scene at that time.

There is not enough room here to present all the evidence to demonstrate that, despite being a scholar of natural history and an administrator, Banks sustained a relationship with quite a number of artists, for whatever reason. Another of them was John Zoffany. He was Banks's prime choice as accompanying artist on Cook's second voyage. When Banks withdrew from that voyage, because there was too little room for his retinue, Zoffany, like Banks, immediately embarked on a different journey and went to Florence, where in August 1772 he began work on his famous "Tribuna" of the Uffizi.

Whitley, in his *Artists and their Friends in England 1700–1799*, quotes from a friendly, almost intimate letter from Zoffany to Banks of 15 January 1774 in

FIG. 6. Sir Joshua Reynolds. Members of the Dilettanti Society, 1777–79. Oil on canvas (one of a pair). Reproduced by permission of The Society of Dilettanti, London.

which he enquires about common friends in London. However, the relations between Banks and Zoffany are enigmatic. The Linnean Society in London treasures a portrait of Dr Solander, which was painted either prior to Cook's second voyage or between 1779 and 1782, just before Solander's death. It is odd that the same figure of Solander is introduced into William Parry's conversation-piece of Banks, Solander and Omai, and the explanation can only be that his is a commemorative painting after Solander's death, which makes use of earlier pictorial sources.

So far one can only state that Zoffany and Banks had many acquaintances in common: they may have been sitters for one, friends for the other, such as the playwright George Colman, or the 4th Earl of Sandwich, represented in a mezzotint by Valentine Green (1774) after a painting by Zoffany (now destroyed). That Zoffany apparently was an excellent painter of flowers and soon after his arrival to England gained the patronage of Lord Bute, one of the great natural history collectors of his age, might point to a closer relationship between Banks and Zoffany — in the field of botany — than meets the eye. It is interesting that when the Royal Gardener at Kew, William Aiton, author of the *Hortus Kewensis* died in early 1793, the botanist Dr Samuel Goodenough, Jonas Dryander, Sir Joseph Banks and John Zoffany were the pall-bearers. It may have been at that time, that is to say after 1792, that Zoffany began work on his neoclassical composition "The Death of Captain Cook", now at the National Maritime Museum. The circumstances which led to the creation of this picture have not been recorded, and the possibility that it was painted with Banks's knowledge or even support is unfortunately not (yet) founded on evidence.

The last relationship between Banks and an English artist of the late 18th century to be recorded here, is that with Paul Sandby. Harold Carter has kindly brought to my attention a letter of 19 December 1773, written by Sandby to Banks from his house in St George's Row, in which Banks is invited to act as godfather for Nancy, Sandby's youngest daughter, who however unfortunately died young.

Half a year earlier Banks and Sandby had gone on a 7-week journey through South and North Wales, which lasted from 25 June to 16 August 1773.

Two years later, in September 1775, Sandby published a series of twelve *Views in South Wales* (Fig. 7), with a dedication to the "Hon. Charles Greville and Joseph Banks Esquire by their ever grateful and much obliged servant Paul Sandby R.A." The story of that trip, which above all was a botanical one, has been well described by F. Hughes (*Burl. Mag.*, July 1975). In recent years more light has been shed on the second name on the dedication page: the Honorable Charles Greville, who is said to have been (like Sir Watkin Williams-Wynn) a pupil of Sandby's. Greville had meant to accompany the travel party, but had to cancel his plans. His contribution to this publication was substantial: he seems to have purchased the formula of the aquatinta process — a French invention by J.B. Le Prince and as yet only rarely practised in England — from the Liverpool engraver Peter O. Burdett, and passed it on to Sandby. This, one assumes, happened sometime in 1775, when Sandby used that technique to great advantage, for in a letter to his Edinburgh pupil John Clerk of Eldin he reports that he has been working on Le Prince's secret: for "these four months past I have scarcely done anything else".

But not only Greville and Sandby were attracted by this new technique. Harold Carter refers to correspondence between Banks and Burdett on the subject in

FIG. 7. Paul Sandby. Chepstow Castle 1773. Published as first plate in Paul Sandby, *XII Views in Aquatinta from Drawings Taken on the spot in South Wales, Dedicated to the Honorable Charles Greville and Joseph Banks, Esquire*, 1775. Aquatint. Reproduced by permission of the National Museum of Wales, Cardiff.

November 1773. Banks obviously looked for a suitable technique in which he could reproduce the drawings for his *Florilegium*. Benjamin Franklin wrote to an unknown correspondent on 3 November 1773: "Mr Banks is at present engaged in preparing to publish the Botanical Discoveries of his voyage, He employs 10 engravers for the plates, in which he is very curious so as not to be quite satisfied in some cases with the expression given by either the graver, Etching or Metzotints, particularly where there is a wooliness or a multitude of small points on a leaf".

If Banks was concerning himself at the end of 1773 with a new and improved medium for printing, a technique that brought more lightness and delicacy to the task than the traditional line engraving, then it was natural that he should have many discussions with his former travel companion and distinguished draughtsman Paul Sandby, a founder of the Royal Academy and father of his god-child. One might speculate as to whether it was Banks's unhappiness with line engraving which suggested the purchase of the formula from Peter Burdett. Certainly, Sandby's *Views in South Wales* is the first aquatinta series of Welsh landscape and monuments to be published.

Probably during the time of cooperation with Sandby, Banks had acquired up to 70 drawings of this master, all of them views around Windsor, and almost all of them painted in watercolour. This was a remarkable acquisition, since, except for natural history drawings, Banks is not known to have purchased a similar large group of drawings at all.

Why Banks acquired these is not clear; friendship may have had a role in it, but it has also been observed that as a former Etonian Banks knew the area around Windsor very well and was indulging in memories. Yet another reason may have been that in the early 1770s, the King developed a liking for Windsor and prepared for structural alterations to make Windsor the seat of the monarchy. Seen in this light, Banks would have enjoyed Sandby's drawings not only from a topographical but also from an architectural and archaeological point of view. Banks retained the Sandby Windsor drawings all his life, probably stored them in a portfolio, where they were safeguarded from light, and bequeathed them to his wife. At her death they were inherited by her family, and in 1876 her descendant Sir W. Knatchbull Bart. sold them at Christie's. Paul Oppé, in his *Drawings of Paul and Thomas Sandby in the Collection of His Majesty the King at Windsor Castle* (London 1947), points out that out of that sale 23 drawings were subsequently bought by the Royal Collection at various times, thus adding to the Windsor drawings which the Royal Collection already held and which had been acquired by George IV, when Prince of Wales.

Two examples of Sandby's Windsor drawings from the Banks collection are the watercolour "View through the Town Gate looking westward down Castle Hill" (1768), a well preserved watercolour drawing which also served as title page to the catalogue of the 1993 Sandby exhibition at Windsor, and "Windsor Castle: The Round Tower, Royal Court and Devil's Tower from the Black Rod", a gouache, which has found its way — through the Felton Bequest — into the National Gallery of Victoria, Melbourne.

This chapter is intended to demonstrate Banks's contacts with the artists of his time. He was familiar with portrait painters of great importance, landscape and figure draughtsmen of distinction, natural history draughtsmen, engravers and printers. Throughout his life Banks had approached the arts from a practical point of view, adopting the role of an employer, promoter and patron, rather than a sensually committed aesthete.

For more than 50 years Banks had used his expertise and discriminating eye in cooperation with artists, employing and patronising them, and often collecting their work. Banks's own area of special concern was botany, but when it came to collecting, there were no firm and exact boundaries. He was certainly interested in portraiture, landscape, topographical scenery, architecture and artefacts. As Farington mentioned in his diary, accuracy of drawing was a principal recommendation to Sir Joseph Banks. Banks's approach to art was sober and practical, and essentially a utilitarian one.

Amongst the family papers in the Kent Archives is this note in his own hand: "I have taken the Lizard, an Animal said to be Endow'd by nature with an instinctive love of Mankind, as my Device, & have Caus'd it to be Engrav'd as my Seal, as a Perpetual Remembrance that a man is never so well Employ'd, as when he is laboring for the advantage of the Public; without the Expectation, the Hope or even a wish to derive advantage of any Kind from the Result of his exertions."

This quote sums him up very neatly. Banks was a public collector, or a collector for the public, and it is pleasing to know that so many of the works which he instigated have survived the ages.

ACKNOWLEDGEMENTS

I wish to acknowledge generous and most helpful support from the Paul Mellon Centre for Studies in British Art, for enabling me to carry out the research which led to this paper. Special thanks are also due to Harold Carter of the Banks Archive and Rex E. R. Banks of the Natural History Museum, London.

R. Desmond, 'The transformation of the Royal Gardens at Kew', in *Sir Joseph Banks: a global perspective* (eds. R.E.R. Banks and others), Royal Botanic Gardens, Kew, 1994, 105–115.

8. THE TRANSFORMATION OF THE ROYAL GARDENS AT KEW

RAY DESMOND

Alexandra Road, Twickenham, Middlesex

Imagine the Royal Botanic Gardens at Kew obliterated by a suburban sprawl of Victorian villas and terraced houses. An alarming thought! Yet it could so easily have happened. It could be argued that it was in no small degree the former Royal Gardens' international renown, so assiduously fostered by Sir Joseph Banks, that persuaded this country's scientific and horticultural community to oppose Treasury's efforts to dismember the Gardens during the 1830s.

Before considering Kew's indebtedness to Banks, it is appropriate to outline the early years of the Royal Gardens. George III amalgamated the two contiguous estates he had inherited at Richmond and Kew — one the property of his grandfather George II, and the other, (horticulturally more important), belonging to his mother, Princess Augusta. The White House at Kew had been leased by her husband, Frederick Prince of Wales in 1730. Few significant changes were made to its garden during Frederick's brief life, but after his death in 1751 his widow with advice from Lord Bute, a knowledgeable botanist and gardener, undertook its improvement. In 1757 William Chambers was commissioned to landscape the grounds. Within an area of some 14 acres an Orangery and a large stove for tender plants were built and a physic garden established under the care of William Aiton, recruited from the Chelsea Physic Garden in 1759. Here garden plants were grown in formal Linnean beds with a decorative display of ornamentals in an adjacent flower garden. A small arboretum to the east of Chambers's Orangery was enriched in 1762 with choice trees transferred from the Twickenham garden of Lord Bute's late uncle, the Duke of Argyll. Bute's friend, John Hill, whose multifarious interests included botany, enjoyed some sort of advisory role at Kew. His *Hortus Kewensis*, published in 1768, listed 50 ferns, between 500 and 600 trees and shrubs and several thousand herbaceous plants then in cultivation in the grounds and glasshouses of the White House. At that time Princess Augusta's garden was, according to Peter Collinson, "the Paradise of our world, where all plants are found, that money or interest can procure".(1) Alexander Garden learnt from John Ellis in 1771, the year in which Cook's *Endeavour* returned home, that "everything curious was now being sent to Kew".(2) So Kew had already acquired an impressive horticultural reputation before Banks appeared on the scene.

According to the *Gentleman's Magazine* for 1793 Banks was about 21 when he first met William Aiton. Three years later, that is in 1767, Banks sent Sydney Parkinson to Kew to draw some of its plants. In August 1771 George III received Banks, now a celebrity after the *Endeavour* voyage, at Richmond.

Lord Bute, once the King's friend and confidant, and now out of favour, lost his consultancy role at Kew when Princess Augusta died in 1772. The young Banks who had impressed George III with his collections of plants and folios of flower drawings by Sydney Parkinson was an opportune replacement for Bute. His collections made on the *Endeavour* voyage together with those accumulated on the earlier Newfoundland trip, confirmed not only Banks's botanical competence but also his enterprise and initiative.

It has not been possible to establish precisely when Banks became involved in Kew's management. Although it was Sir John Pringle, shortly to be elected President of the Royal Society, who had suggested to the King in 1772 that a plant collector should be despatched to the Cape, the prompting probably came from Banks who had briefly visited South Africa on the *Endeavour*'s return voyage. Certainly by 1773 Banks commanded some authority at Kew but never operated in any official capacity. In a letter to the Spanish ambassador in 1796 he modestly described himself as exercising "a kind of superintendence over his [Majesty's] Royal Botanic Gardens",[(3)] William Townsend Aiton, who succeeded his father at Kew in 1793 acknowledged this status when he told a correspondent in 1801 that "This establishment is placed under the direction of Sir Joseph Banks".

Since only a few letters between Banks and the King have survived, it is not unreasonable to suppose that his communications with George III were usually in person. We know they frequently met at Kew, often on Saturdays when matters relating to the Royal Gardens were discussed during a routine tour of inspection. George III's respect and regard for Banks endured until the King's final breakdown in 1810.

The combined skills and efforts of Lord Bute, John Hill and William Aiton had converted Kew into one of the best-stocked gardens in the country. Banks, however, soon perceived that if it were to compete with rival institutions in Paris and Vienna then the prevailing indiscriminate acquisition of plants would have to be replaced by more precisely defined collecting objectives. He envisaged Kew eventually becoming a botanical garden participating in the receipt, acclimatization and despatch of plants within an expanding British Empire. The earliest use of the descriptive phrase, "The Royal Botanic Gardens at Kew" that I have found appears in an article contributed by Francis Masson to the *Philosophical Transactions of the Royal Society* in 1776. The epithet "Royal Gardens" continued in use but the more frequent insertion of the word "Botanic" confirmed Kew's changing role.

Francis Masson, an "under-gardener" at Kew in his early thirties, became the Royal Gardens' first collector following Sir John Pringle's submission to the King in 1772. During his three years' residence at the Cape he made several expeditions in the interior, finding new species of Ericas, Pelargoniums, Stapelias and much else, making Kew, so Banks boasted, superior to all other European botanical gardens. After several years collecting in the Canaries, Azores and the West Indies, Masson returned to South Africa in 1786, remaining there for the next nine years. When Banks sought permission of the Spanish Ambassador in 1796 for a Kew gardener to be stationed in Buenos Aires he had Masson in mind for the post. His overture unsuccessful, Masson left two years later for North America where he died in 1805, still collecting for Kew.

During the mid-1780s Banks had intended sending Masson to Botany Bay in Australia whose flora was sparsely represented in the Royal Gardens by the few specimens brought back on Captain Cook's three voyages. Two Kew-trained gardeners who joined H.M.S. *Guardian* in 1789 on its ill-fated voyage to the new colony, were never heard of again after its shipwreck. Some years later two more Kew collectors, Allan Cunningham and Peter Good, were also to die in the Antipodes. Disease was a constant hazard for these early collectors: David Nelson cast adrift with Captain Bligh after the *Bounty* mutiny died of fever on Timor, and William Kerr succumbed after only two years in Ceylon.

Sir Joseph Banks had identified three floristic regions of particular interest to Kew: Australia, the Cape and South America. Having failed in 1796 to place a collector at Buenos Aires, he tried again in 1802, stressing Kew's importance as a royal residence by reminding the Foreign Secretary that a new palace for George III was being built there. Again he was frustrated by Spain's difficulties with its American colonies, and William Kerr, the Kew gardener who had been destined for South America, found himself instead in 1803 on a ship bound for China.

Here we have a typical example of Banks's manipulation of his network of connections, in this instance, the East India Company. At his request the Company agreed to defray the cost of maintaining Kew's collector at its station in China and to bring back, free of charge, on their East Indiamen all the plants he acquired.

Banks demanded more than the mere acquisition of plants from his collectors. Kerr, for example, was instructed to inform Aiton about climate and soils, local medicinal and culinary uses of the vegetation, techniques for cultivating miniature shrubs, fibres selected for cordage and the manuring of crops. Since the xenophobic Chinese restricted the movement of all foreigners to the ports of Canton and Macao, Kerr was handicapped in his horticultural investigations and limited to the plants he could obtain in local nurseries and gardens. Nevertheless he introduced to this country outstanding horticultural novelties such as the double yellow Banksian rose, *Pieris japonica*, a variety of *Lonicera japonica*, the tiger Lily, and a yellow shrub, later renamed *Kerria japonica* as a tribute to his achievements.

When Banks visited the King at Windsor Castle in 1810 for the last time, he informed him that William Kerr had been promoted to Superintendent and Chief Gardener of the new Botanic Garden in Ceylon. This appointment which rewarded Kerr's diligence and dedication in China was not take up until 1812. Banks had great expectations of his superintendency: "an abundance of mountain plants such as will thrive in our greenhouses, if not in the open air", he told the botanist Nicolaus Thomas Host, "in truth our hothouses overflow at present so much with intertropical plants, that I scarce wish for additions of that kind".(4) When Kerr died just two years later, another Kew gardener, Alexander Moon, succeeded him.

Because the disruption of shipping caused by the Napoleonic war impeded the safe transportation of living plants, Kerr became Kew's last official collector until the cessation of hostilities in 1814. With the return of peace William Townsend Aiton immediately sought Banks's consent to resume collecting, assuring him that he had at his disposal "men of sound principles and invaluable zeal for the service, having the best requisites of knowledge, and desire to offer themselves as collectors, and who will perform this duty in any part of the world".(5) Banks unhesitatingly agreed: "the connection I have been permitted to form with the Royal Garden at Kew is among those most grateful to my feelings, and I beg you to be assured that as long as I shall be permitted to continue it, I shall cherish and improve it to the best of my power".[6] Determined to re-establish Kew's supremacy, he judged the imperial gardens at Schönbrunn in Vienna to be the only serious competitor. He favoured the Cape and New South Wales whose flora did not require hothouses generating extreme heat and humidity.

With the Prince Regent's permission and Treasury's backing, two Kew gardeners — Allan Cunningham and James Bowie — sailed for Rio de Janeiro in October 1814 to await a passage to their respective destinations. The two years they spent collecting in the neighbourhood of Rio, the nearby Organ mountains and São Paulo, some 400 miles distant, went some way to fulfilling Banks's ambition to extend the range of the South American flora at Kew. Both men left Brazil in September 1816, Bowie to the Cape and Cunningham to Botany Bay. The last letter they received from Banks before their departure reminded them that "one plant from the Cape of Good Hope or from N.S. Wales that will live in the dry stove or perhaps in the greenhouse is worth in Kew Gardens a score of tender plants that require the hot house or roaster".

Before he left Brazil, Cunningham renewed his acquaintance with James Hooper, a foreman gardener at Kew, on board H.M.S. *Alceste* taking Lord Amherst on a diplomatic mission to China. Hooper's contract required him to assist Dr Clarke Abel, the mission's physician, in the collecting and naming of plants and in sowing seeds in pots for despatch to England. Banks seldom missed any opportunity to take advantage of missions and expeditions. In the same year, i.e. 1816, another Kewite, David Lockhart, was seconded as assistant gardener to the botanist Christen Smith on Captain Tuckey's ill-fated survey of the Congo River. It was Banks who nominated Archibald Menzies as surgeon and naturalist on Captain Vancouver's survey of the north-west coast of America in 1791–95, and provided him with a small greenhouse to house the plants he was expected to collect for Kew.

When the government wanted a naturalist to join its survey of the west coast of Africa in 1785 it automatically turned to Banks who urged William Aiton to find a suitable candidate: "the opportunity is not to be lost", he wrote, "we shall get specimens for me and seeds for Kew Gardens ... For God's sake be active and do not let such an opportunity slip you".[7]

Without doubt the most rewarding voyage for Kew was that of H.M.S. *Providence* under Captain Bligh whose two-fold mission was to bring the breadfruit from the Pacific to the West Indies and plants to the Royal Gardens. When the ship docked at London in 1793 it had on board over 2000 healthy plants from Timor, Tahiti, St Vincent and Jamaica for Kew.

Three years later a comparable collection from the Botanic Garden at Calcutta reached Kew under the care of the Kew gardener, Peter Good. William Roxburgh, Superintendent at the Calcutta garden and, more especially, his successor Nathaniel Wallich, frequently donated plants and seeds from South Asia and the East Indies. The services that Banks rendered to the newly formed colonial botanical gardens, sometimes managed by former Kew staff, were gratefully reciprocated with consignments for the Royal Gardens: from Alexander Moon at Peradeniya in Ceylon, Alexander Anderson and George Caley at St Vincent, David Lockhart at Trinidad, and Hinton East and James Wiles at Jamaica.

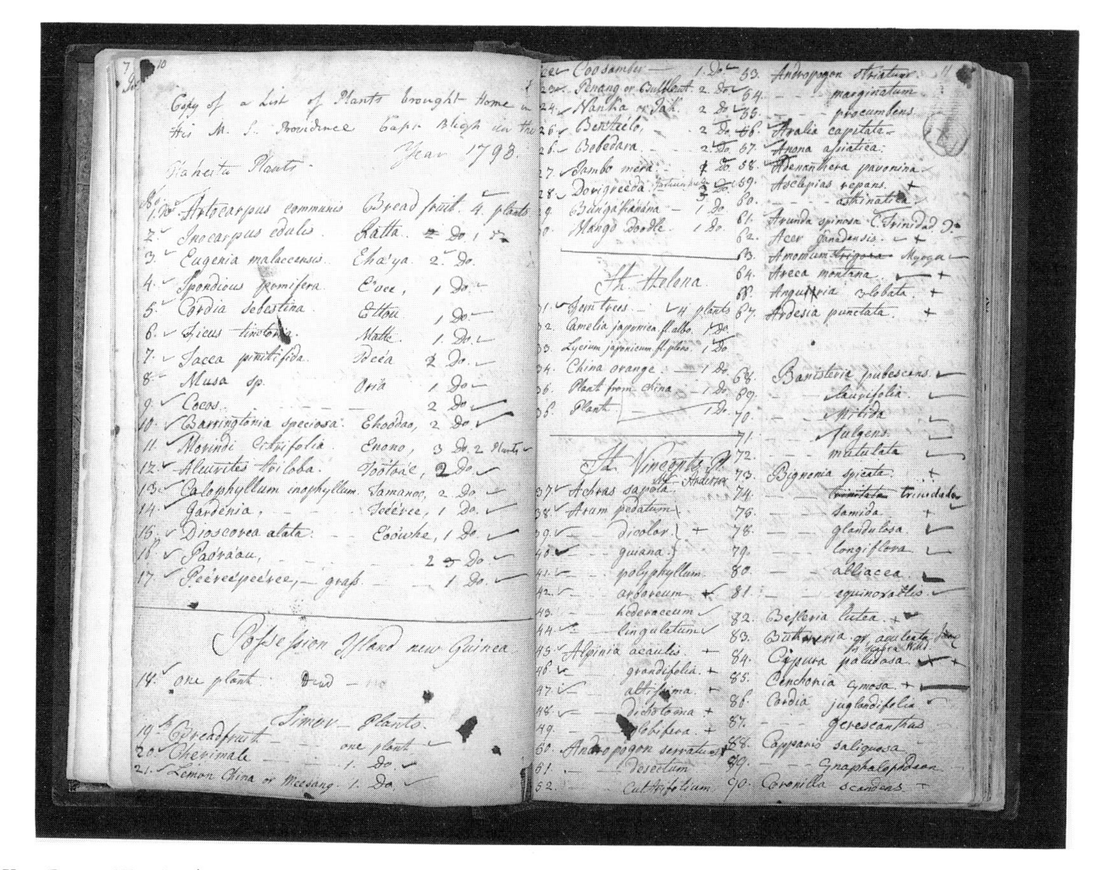

Copy of a List of Plants brought Home in
His M. S. Providence Capt. Bligh in the
Year 1793.

Taheitu Plants

1. Artocarpus communis — Bread fruit. 4. plants
2. Inocarpus edulis — Ratta. 1
3. Eugenia malaccensis — Eha'ya 2 Do.
4. Spondious pomifera — E'vee, 1 Do.
5. Cordia sebestina — Ettou 1 Do.
6. Ficus tinctoria — Matte 1 Do.
7. Tacca pinnatifida — Peea 2 Do.
8. Musa sp. — Ora 1 Do.
9. Cocos — 2 Do.
10. Barringtonia speciosa — Ehoodaa, 2 Do.
11. Morinda citrifolia — Enono, 3 Do. 2 Plants
12. Aleurites triloba — Tootooe, 2 Do.
13. Calophyllum inophyllum — Tamanoo, 2 Do.
14. Gardenia, — Teeairee, 1 Do.
15. Dioscorea alata — Eoowhe, 1 Do.
16. Paoraau, — 2 Do.
17. Peereepeeree, — grass — 1 Do.

Possession Island new Guinea

14. one plant

Timor Plants

19. Breadfruit — one plant
20. Cherimale — 1. Do.
21. Lemon China or Meesang. 1. Do.

22. Coosamber — 1. Do.
23. Pinang or [illegible] 2 Do.
24. Nanka or Jak 2 Do.
25. Benteelo, 2 Do.
26. Belledara 2 Do.
27. Jambo mera 4 Do.
28. Dovigreeda 3 Do.
29. Bunga Kanana 1 Do.
30. Mango Doodle 1 Do.

St. Helena

31. Gum trees — 4 plants
32. Camelia japonica fl. albo 1 Do.
33. Lycium japonicum fl. pleno 1 Do.
34. China orange — 1 Do.
35. Plant from China — 1 Do.
36. Plant — 1 Do.

St. Vincents

37. Achras sapota
38. Arum pedatum
39. — bicolor
40. — guiana
41. — polyphyllum
42. — arboreum
43. — hederaceum
44. — lingulatum
45. Alpinia acaulis
46. — grandifolia
47. — altissima
48. — dichotoma
49. — [illegible]
50. Andropogon serratus
51. — desertum
52. — [illegible]folium

53. Andropogon striatus
54. — marginatum
55. — procumbens
56. Aralia capitata
57. Anona asiatica
58. Renanthera pavonina
59. Asclepias repens
60. — asthmatica
61. Arundo spinosa Trinidad
62. Acer [illegible]
63. Amomum Myoga
64. Areca montana
65. Angustura trilobata
66. Ardesia punctata
68. Banisteria pubescens
69. — laurifolia
70. — nitida
71. — fulgens
72. — mutilata
73. Bignonia spicata
74. — trinitata
76. — Samida
78. — glandulosa
79. — longiflora
80. — alliacea
81. — equinoxalis
82. Besleria lutea
83. Buttneria aculeata
84. Cyperus paludosa
85. Conchonia ramosa
86. Cordia juglandifolia
87. — [illegible]
88. Capparis saligna
89. — gnaphalodes
90. Coronilla scandens

Kew Record Book, 1793–1809. The entry which lists the plants brought to Kew Gardens by Captain Bligh on H.M.S. *Providence* in 1793.

Exchanges of plants and seeds were also made with individuals like the French botanist, Dumont de Courset and with institutions like the Imperial Gardens at Schönbrunn.

It was, however, to Kew's own collectors that the Royal Gardens were most indebted. They were usually recruited from those gardeners who demonstrated a combination of horticultural skills, botanical aptitude, and a knowledge of foreign vegetation. Banks refused to employ George Caley as an official collector in New South Wales until he was familiar with all the Australian plants in cultivation at Kew. "It is important for you to distinguish those that are new in order to send home new plants and leave those already sent home behind, a knowledge indispensably necessary for a botanic collector".[8] Personal qualities were also taken into account. Peter Good was chosen for his "good sense and daring character", the selection of Allan Cunningham and James Bowie was influenced by their "Honesty, Sobriety, Diligence, Activity, Humility and Civility".[9] Banks had a predilection for the Scots in whom he discerned "the habits of industry, attention and frugality". Bachelors were preferred for a career that entailed years of solitary service abroad.

No collector ever undertook a mission without a thorough briefing from Banks. The instructions he gave Archibald Menzies are typical of the indocrination they all received. Menzies was required to note the vernacular as well as the scientific names of the plants he collected or observed; his packets of seeds were to be labelled with details of soil, climate and other relevant cultivation data for the benefit of Kew's propagators; whenever he had doubts about Kew's ability to germinate certain seeds, the plants themselves were to be potted and placed in the ship's glasshouse. Menzies and all other collectors were strictly forbidden to give any plants, seeds or cuttings to anyone else, especially nurserymen. Banks was determined that His Majesty's Gardens should boast an unrivalled collection.

But the skills and the dedication of these collectors were no guarantee that their plants would arrive at Kew safely and in good health. Until the use in the late 1830s of a sealed glass container known as a Wardian case, long sea voyages which could exceed six months for ships sailing from India and beyond were often fatal for plant cargoes. The casualty rates were so high that in the 1790s Kew installed an 'infirmary', as Banks called it, for any sickly plants that had survived the voyage. Despite exhortation from Banks, ships' crews, understandably reluctant to tend tubs of plants cluttering the deck, left them unprotected during storms and neglected to ensure they received adequate watering. The captain of one ship leaving Sydney in 1789 was requested by Banks to ensure that "if any other plants except those intended for the King be taken on board, no water shall be issued to them until the King's plants shall have had their full allowance". Two young gardeners, James Smith and George Austin, who had charge of the glasshouse on H.M.S. *Guardian* bound for Australia were reminded by Banks of the deleterious effects of salt sea spray, the need for appropriate watering, light and air, and the constant threat from rats, dogs, cats and the miscellaneous livestock ships usually had on board. Above all they were warned to "beware of liquor, as one drunken bout may render the whole of your care during the voyage useless, and put your character in a very questionable situation."[10] At least one of the ships from China employed a Chinese gardener to look after those plants destined for the Royal Gardens. On

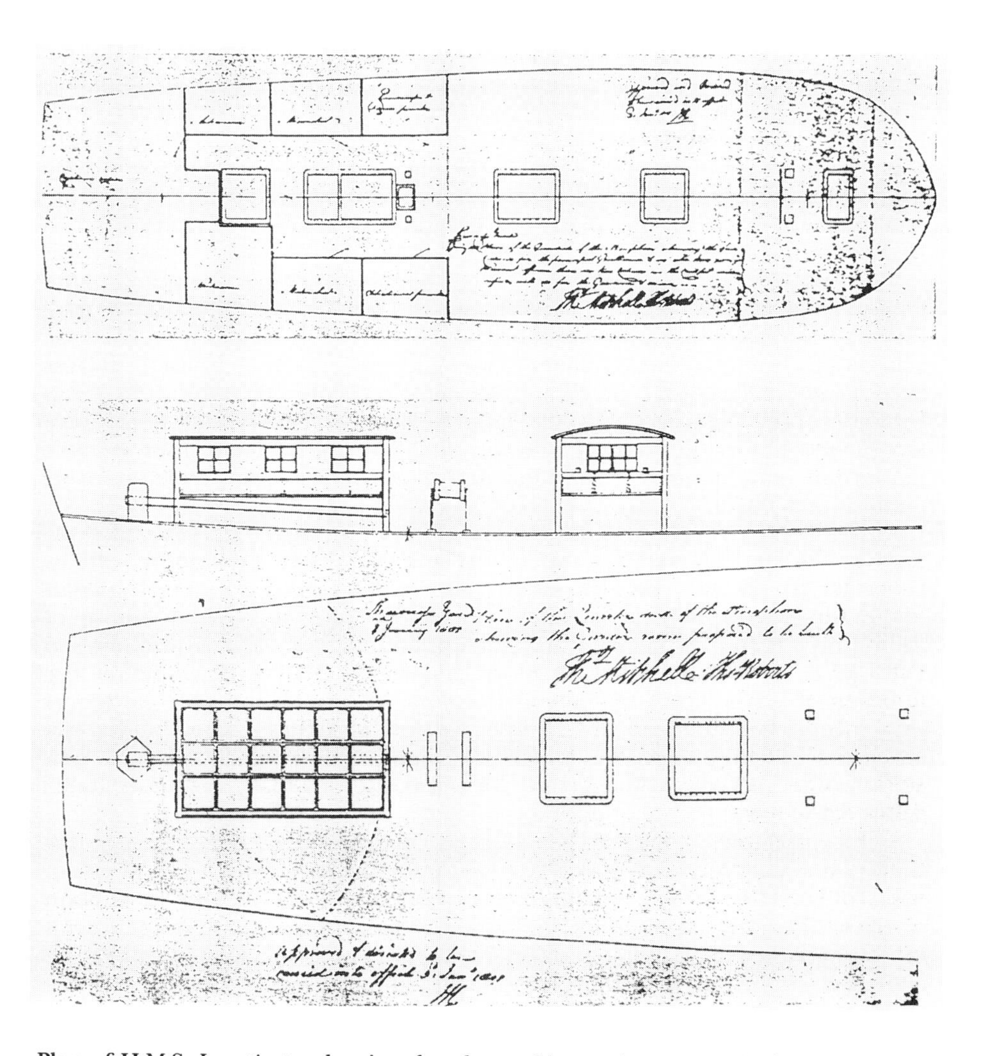

Plan of H.M.S. *Investigator* showing the plant cabin on the quarter deck. (R.A. Austin *Matthew Flinders on the Victorian Coast* 1794, p. 17).

arrival in England, he was accommodated at Kew and given the opportunity to learn something of British gardening techniques before returning East escorting another consignment of plants. Banks believed that plants were best protected in a small glasshouse, variously known as a 'plant cabin', 'garden hutch' or 'botanic conservatory'. He himself supervised their installation on the quarter deck of several ships including H.M.S. *Investigator* which took his botanical assistant, Robert Brown, to Australia in 1801.

Sufficient plants survived the hazards of transportation to justify additional glasshouses at Kew: one for Cape plants in 1788, one for the Australian flora in 1792 and yet another one in 1802. "Kew Gardens proceeds with increased vigor", Banks informed the Swedish naturalist, Retzius, in 1792. "The additions of plants lately received are indeed very interesting. We have three Magnolias from China, one only of which was before known among us and that only from Kaempfer ... Epidendrums blossom daily and Vanilla is as high as the glass and will soon produce flowers". By 1817, 837 feet of glass were in use while the population of hardy plants had increased spectacularly.

A published list of the plants in cultivation at Kew had long been a desideratum. Sir John Hill had enumerated 3,400 species in his *Hortus Kewensis* (1768). Appropriating the same title, William Aiton was the putative author of the new catalogue, published in three volumes in 1789. Its dedication to the King revealed that Aiton had devoted more than 16 years to its compilation. His terse acknowledgement to "the assistance of men more learned than himself" did less than justice to the extent of his indebtedness to Daniel Solander and Jonas Dryander, both botanical assistants employed by Banks. The diagnoses were laregly Solander's; after his death in 1782, Dryander, curator of Banks's herbarium and library, continued the compilation and supervised its publication.

In April 1803, Banks proposed a revision of the *Hortus Kewensis*, tactfully urging William Townsend Aiton to consult Dryander about its feasibility. Dryander, who readily concurred, had completed the first two volumes and part of the third volume before his death in October 1810. With the co-operation of Robert Brown, Banks's latest curator, the work was completed in five volumes in 1813. W.T. Aiton acknowledged more generously than had his father the indispensable assistance of his collaborators. The following year gardeners were offered a condensed and cheaper version, an *Epitome*, which added 300 more species with a selection of vegetables and fruits cultivated at Kew. A new edition of this *Epitome*, almost ready for publication in 1830, was unavoidably delayed and eventually abandoned.

Sir Joseph Banks valued the *Hortus Kewensis* both as a work of reference and as a public declaration of Kew's achievements. James Bowie, newly arrived in South Africa in 1817, learnt that his first task was to collect replacements for the Masson plants which had subsequently died at Kew: "their names stand in the *Hortus Kewensis*; it is therefore most important to the credit of the Garden to replace the plants and make the Garden correspond with the Catalogue".

The *Hortus Kewensis* for 1789 listed about 5,600 species which expanded to 11,000 in the second edition. As one would expect, hardy plants predominated; in 1789 there were just over 2000 British, 1400 European and nearly 700 North American species in cultivation. The availability of glasshouses obviously controlled the population of tender plants; the South African flora represented the largest group with more than 700 species, a third of which were Masson introductions. In the 1810–13 edition of the *Hortus Kewensis* British species showed a slight decline; the number of Cape plants doubled, and Australian plants shot up to over 300, thanks largely to the efforts of George Caley and also to Robert Brown and Peter Good on H.M.S. *Investigator*.

The *Hortus Kewensis* encouraged Banks to consider a more ambitious project — a publication illustrating a selection of the plants it listed. Sir John Hill had published *Twenty-five new plants rais'd in the Royal Garden at Kew* in 1773 and Kew's principal rival, the Imperial Gardens in Vienna, had depicted 300 of its exotics in *Hortus Botanicus Vindobonensis* (1770–76). In January 1789 the first part

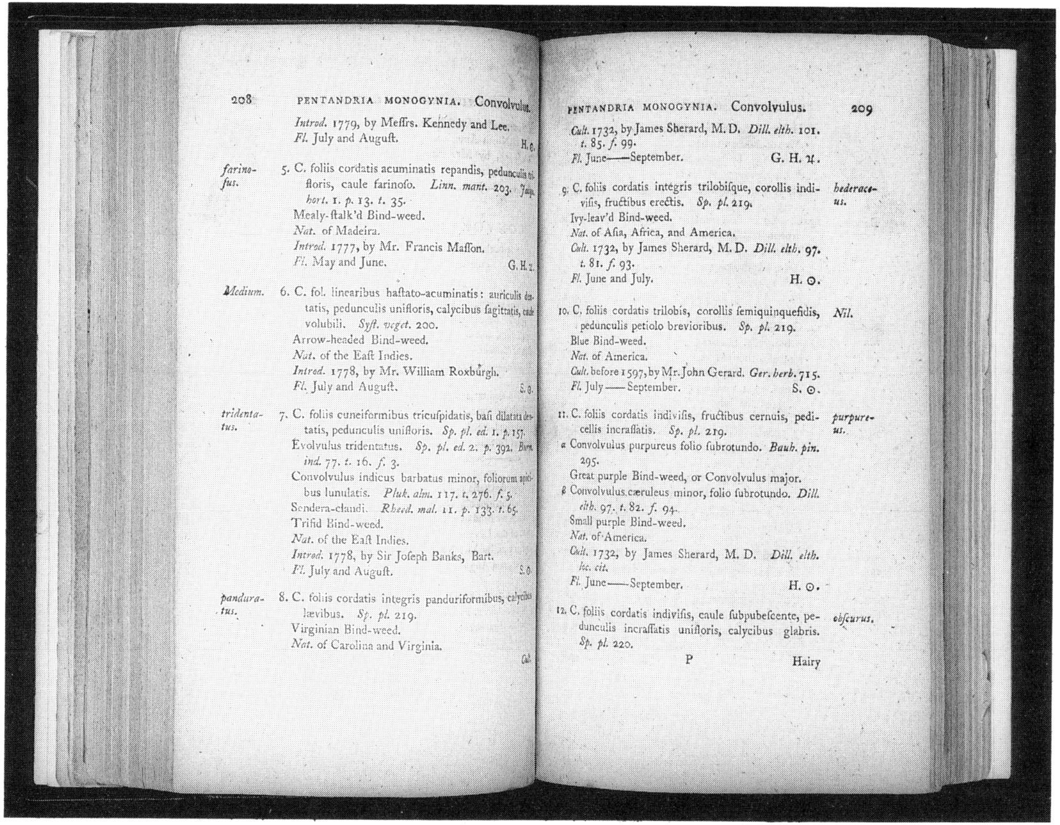

208 PENTANDRIA MONOGYNIA. Convolvulus.

Introd. 1779, by Meſſrs. Kennedy and Lee.
Fl. July and Auguſt. H. ⊙.

farinoſus. 5. C. foliis cordatis acuminatis repandis, pedunculis triſloris, caule farinoſo. *Linn. mant.* 203. *Jacq. hort.* 1. *p.* 13. *t.* 35.
Mealy-ſtalk'd Bind-weed.
Nat. of Madeira.
Introd. 1777, by Mr. Francis Maſſon.
Fl. May and June. G. H. ♃.

Medium. 6. C. fol. linearibus haſtato-acuminatis: auriculis dentatis, pedunculis unifloris, calycibus ſagittatis, caule volubili. *Syſt. veget.* 200.
Arrow-headed Bind-weed.
Nat. of the Eaſt Indies.
Introd. 1778, by Mr. William Roxburgh.
Fl. July and Auguſt. S. ⊙.

tridentatus. 7. C. foliis cuneiformibus tricuſpidatis, baſi dilatata dentatis, pedunculis unifloris. *Sp. pl. ed.* 1. *p.* 157.
Evolvulus tridentatus. *Sp. pl. ed.* 2. *p.* 392. *Burm. ind.* 77. *t.* 16. *f.* 3.
Convolvulus indicus barbatus minor, foliorum apicibus lunulatis. *Pluk. alm.* 117. *t.* 276. *f.* 5.
Sendera-clandi. *Rheed. mal.* 11. *p.* 133. *t.* 65.
Trifid Bind-weed.
Nat. of the Eaſt Indies.
Introd. 1778, by Sir Joſeph Banks, Bart.
Fl. July and Auguſt. S. ⊙.

panduratus. 8. C. foliis cordatis integris panduriformibus, calycibus lævibus. *Sp. pl.* 219.
Virginian Bind-weed.
Nat. of Carolina and Virginia.

Cult.

PENTANDRIA MONOGYNIA. Convolvulus. 209

Cult. 1732, by James Sherard, M. D. *Dill. elth.* 101. *t.* 85. *f.* 99.
Fl. June——September. G. H. ♃.

hederaceus. 9. C. foliis cordatis integris trilobiſque, corollis indiviſis, fructibus erectis. *Sp. pl.* 219.
Ivy-leav'd Bind-weed.
Nat. of Aſia, Africa, and America.
Cult. 1732, by James Sherard, M. D. *Dill. elth.* 97. *t.* 81. *f.* 93.
Fl. June and July. H. ⊙.

Nil. 10. C. foliis cordatis trilobis, corollis ſemiquinquefidis, pedunculis petiolo brevioribus. *Sp. pl.* 219.
Blue Bind-weed.
Nat. of America.
Cult. before 1597, by Mr. John Gerard. *Ger. herb.* 715.
Fl. July——September. S. ⊙.

purpureus. 11. C. foliis cordatis indiviſis, fructibus cernuis, pedicellis incraſſatis. *Sp. pl.* 219.
α Convolvulus purpureus folio ſubrotundo. *Bauh. pin.* 295.
Great purple Bind-weed, or Convolvulus major.
β Convolvulus cæruleus minor, folio ſubrotundo. *Dill. elth.* 97. *t.* 82. *f.* 94.
Small purple Bind-weed.
Nat. of America.
Cult. 1732, by James Sherard, M. D. *Dill. elth. loc. cit.*
Fl. June——September. H. ⊙.

obſcurus. 12. C. foliis cordatis indiviſis, caule ſubpubeſcente, pedunculis incraſſatis unifloris, calycibus glabris. *Sp. pl.* 220.

P Hairy

W. Aiton *Hortus Kewensis* 1789.

of L'Heritier's *Sertum Anglicum* appeared with engravings of plants found in London's gardens including the Royal Gardens. That Kew's treasures should first be illustrated in a foreign publication must surely have irritated Banks. He had already decided to launch a periodical, a preference probably influenced by the *Botanical Magazine*, a monthly illustrated journal, started by William Curtis in February 1787. The regular services of a botanical artist were essential for the success of such a project. Banks had always believed that any botanical garden worthy of its name should employ a resident artist to record choice and rare plants in its collection. So when Francis Bauer, an Austrian artist trained in the Imperial Gardens in Vienna, came to London in 1789, he was enticed to Kew, Banks paying his salary from his own pocket. Probably early in 1790 Bauer began his half century's service at Kew while everyone eagerly awaited the first number of this new periodical. What eventually appeared were the three parts of *Delineations of Exotic Plants cultivated in the Royal Gardens at Kew* between 1796 and 1803. Because this splendid folio, depicting 30 Ericas collected by Francis Masson in the Cape, failed to recoup its production costs, the periodical proceeded no further. With an annuity from Banks, Bauer remained at Kew until his death in 1840.

Despite his achievements at Kew, Banks was not without his critics. Dean William Herbert deplored "the narrow-minded doctrine of Sir Joseph Banks, that he could render the King's collection superior to others by monopolising its contents".[11] His charge that Kew showed great reluctance to part with any of its plants, was a complaint that would have been whole-heartedly endorsed by the Marquis of Blandford who constantly pestered Aiton for specimens and cuttings for his garden at Whiteknights. Under pressure Banks agreed to the Marquis's importunity provided "the superiority which his Majesty's Garden has for some years held over the other Gardens of his country is not put in hazard by parting with too many species which Kew alone possesses".[12] Here we have the nub of the matter: Banks's constant concern to preserve the uniqueness of Kew's collection. It should, however, be mentioned that Frederick Scheer, a resident of Kew parish, claimed it was Banks's policy to permit the distribution of rare plants after they had flowered. Even Banks had to concede that donations of plants to the crowned heads of Europe was a legitimate function of a Royal Garden; nor did he object to other botanical gardens as recipients; and he willingly supplied economic plants to British colonies. But my examination of the records reveal that not many individuals were beneficiaries of Kew's generosity. So perhaps there was some justification for Dean Herbert's complaint.

But in fairness to Banks, it must be stated that he usually restrained his own collecting instincts to Kew's advantage, redirecting there any new exotics delivered personally to him. He envied the expenditure that royal patronage had lavished on the Schönbrunn garden in Vienna while Kew was managed with "well considered economy". He frequently used his influential connections to acquire plants at no cost to Kew. "I trust, good sir", he tactfully enquired of Governor Hunter in New South Wales, "that when you make your excursions, or when you send parties into new districts, you will not forget that Kew Gardens is the first in Europe, and that its Royal Master and Mistress never fail to receive personal satisfaction from every plant introduced there from foreign parts when it comes to perfection".[13]

He encouraged loyalty to Kew and a competitive spirit in all those who collected at his behest. Dr Clarke Abel, the physician and naturalist accompanying the British Embassy to China, was reminded that "it is desirable that as many of the new plants as possible should make their first appearance at the Royal Gardens".[14] When Banks discovered that the French were fitting out a vessel to explore the north and west coasts of Australia, he lost no time ordering Allan Cunningham to join a British warship about to undertake a similar voyage. "This will give you an opportunity of collecting plants which could by no other means be obtained and of enriching the Royal Gardens at Kew with plants which otherwise would have been added to the Royal Gardens at Paris, and have tended to render their collections superior to ours".[15]

To summarize briefly: the Banks era through an intensification of collecting programmes significantly enriched the Royal Gardens, transforming them into an important centre for the global transfer of plants; links were forged with overseas botanical gardens; the living collections in the Royal Gardens were generously made available to all botanists for the purpose of research; in Victoria's reign the Hooker dynasty at Kew were inspired by Banks's vision; and in the superb drawings of Francis Bauer we have ample evidence of the floral riches that made Kew such an oustanding garden.

Notes

(1) P. Collinson, 21 August 1766, in W. Darlington, *Memorials of John Bartram and Humphry Marshall*, 1844, 282.

(2) J. Ellis, 2 January 1771, in J.E. Smith, *Selection from Correspondence of Linnaeus*, vol. 1, 1821, 583

(3) J. Banks, Dawson Turner transcripts, Natural History Museum, London, [hereafter DT], vol. 10(1), fol. 38–43.

(4) J. Banks, 25 Feb. 1812, DT, vol. 18, fol. 46–7.

(5) W.T. Aiton, 29 May 1814, DT, vol. 19, fol. 36–7.

(6) J. Banks, 7 June 1814, DT, vol. 19, fol. 40–41.

(7) J. Banks, 29 August 1785, Hyde Collection, Somerville, New Jersey.

(8) J. Banks, 27 August 1798, Banks papers, Mitchell library, Sidney.

(9) J. Banks, 18 Sept. 1814, Kew Collectors: Cunningham and Bowie, Kew Archives, fol. 15.

(10) J. Banks, July 1789, DT, vol. 6, fol. 196–203.

(11) W. Herbert, *Amaryllidaceae*, London, 1837, 247–8.

(12) J. Banks, 4 Dec. 1797, DT, vol. 10(1), fol. 214–6.

(13) J. Banks, 30 March 1797, DT, vol. 10(2), fol. 93–5.

(14) J. Banks, 10 Feb. 1816, DT, vol. 19, fol. 240.

(15) J. Banks, 13 Feb. 1817, DT, vol. 20, fol. 17–19.

B. Elliott, 'The Promotion of Horticulture', in *Sir Joseph Banks: a global perspective* (eds. R.E.R. Banks and others), Royal Botanic Gardens, Kew, 1994, 117–131.

9. THE PROMOTION OF HORTICULTURE

BRENT ELLIOTT

The Lindley Library, The Royal Horticultural Society, London

Introduction

The position of Sir Joseph Banks in the history of horticulture derives less from his own gardening, or from any contributions he made in his own person, important though they were, than from his role in stimulating others, and especially in founding the Horticultural Society of London, which later became the Royal Horticultural Society.

The nineteenth century saw an astonishing efflorescence of horticulture, among the most notable features of which were the development of modern greenhouse cultivation, the beginnings of modern plant pathology and pest control, experiments in the acclimatisation of exotic plants and the first systematic programmes of plant hybridisation. In all these fields, Banks played an instigating role, either through his own activities, or through his patronage and encouragement of others.

The early Victorians tended to look back on their predecessors — those of Sir Joseph Banks's era — as by and large a benighted lot. "Only four years before the first Reform Bill", said Donald Beaton, the leading gardening journalist of the mid-century, "some of the best gardeners in the country did not know or understand the principle of potting plants".[(1)] When such men pointed reverentially to predecessors who could be said to have put gardening on the right path, it was predominantly John Claudius Loudon, Thomas Andrew Knight, and John Lindley whom they named; if Banks's name appeared in such a context, it was usually only in his capacity as a founder of the Royal Horticultural Society. Important as that role was, however: one should note that Knight was virtually Banks's protege, that Lindley started his career as a botanist by working as an assistant in Banks's library, and that while there is no indication of a personal connection between Banks and Loudon, Loudon did at least use the Banks herbarium in his own researches, and he readily acknowledged his debt to Knight. So, although he has seldom been recognised in this role, Banks can already be seen looming behind the founders of 19th-century horticulture.

In what follows, I am going to deal with Banks purely in his English, or British, context. His role in stimulating horticultural development on the continent is a subject still to be investigated. I will note here only his role in procuring the services of John Graefer as a head gardener for the King of Naples at Caserta, and remark that, from Graefer's umbrageous comments in his correspondence with Banks about the feeble state of the Italian botanic gardens, it is apparent that he regarded Italian horticulture in rather the same light that the Victorians regarded the English horticulture of his day.[2]

Banks's own contributions

Let me begin by examining what Banks could be said to have contributed to horticulture in his own person.

a) Plant introductions

First and most obviously, there is his record of achievement as a plant collector. Nearly 7000 species of plants were introduced into cultivation in this country during the reign of George III, mainly through the efforts of the collectors Banks sent out under the auspices of Kew. While this figure is a small trickle by comparison with the floods of plants that arrived after the invention of the Wardian case in the 1820s, Banks's achievement in coordinating the greatest quantity of introductions ever previously seen should be recognised. And, of course, two plant names in particular record our gardens' debt to him: *Rosa banksiae*, which Kerr introduced in 1807, and *Hydrangea* 'Sir Joseph Banks', the currently accepted name for the plant Banks introduced in 1788, and which was originally diagnosed as a species under the name *Hydrangea hortensia.*

The gardening press of the early 19th century, however, shows, if not an animus against, at least a reserve towards Banks and attributes introductions, where possible, to the individual collectors. While Banks's name occurs frequently in the descriptions published in the *Botanical Magazine* and the *Botanical Register*, it is usually his herbarium that is referred to, as a source of diagnostic information; and in practice, that could often be taken to mean Robert Brown. Only 10 of the plants illustrated in the *Botanical Magazine* during Banks's lifetime came from Kew, and fewer still from his private garden at Spring Grove; only three Spring Grove plants were depicted in the *Botanical Register* (plus a fourth for which Banks had provided a drawing). Other proprietors, like Aylmer Bourke Lambert, were by contrast praised for their liberality with plants.[3]

After Banks's death, this reserve modulated into open attack. William Herbert, in his *Amaryllidaceae*, published in 1837, criticised Banks for being secretive about his plants. "The illiberal system established at Kew Gardens by Sir Joseph Banks, whereby the rare plants collected there were hoarded with the most niggard jealousy, and kept as much as possible out of the sight of any inquirer, led in the first instance to a feeling of satisfaction, whenever it was known that the garden had been plundered, and some of its hidden treasures brought into circulation. It was the narrow-minded doctrine of Sir J. Banks, that he could only render the King's collection superior to others by monopolizing its contents; and by doing so he rendered it hateful and contemptible: whereas, if he had freely given and freely received, and made its contents easily accessible to those who were interested in them, it would have been a pleasure and a pride to the nation." (Herbert acknowledged that R.A. Salisbury, with Banks one of the founders of the Horticultural Society, had remonstrated with him about this closeness - in other words, that the Society was not to be associated with its founder's behaviour).[4]

It should be noted, however, that on the occasions when Banks did present a plant of his own for illustration in the press, it was often to make an important point about horticultural practice. When the *Botanical Register* illustrated a tree mignonette, it also reported Banks's investigations into French methods of training it; [5] a myrtle-leaved orange tree from Spring Grove was accompanied by his recommendations for grafting it.[6] Most importantly of all, when describing "Sir Joseph Banks's Aerides" (*Aerides paniculatum*) in 1817, it reported Banks's invention of the orchid basket as a means of growing epiphytic orchids.[7] This was a major breakthrough in orchid culture, and although the cultivation of epiphytes did not become properly successful until gardeners realised that they were not parasitic, Banks's technique underlay all subsequent developments in the field.

b) The rockery at the Chelsea Physic Garden

A second claim for Banks is his contribution to the first rock garden in the country, at the Chelsea Physic Garden.

The legend has it that Banks brought back from Iceland, as a geological curiosity, a collection of lava from Mount Hekla, and presented it to the Apothecaries' Garden, where it was used to make the rock garden. It has generally been assumed that this rockery would have been a rather rococo production, in which the interest of the geological specimens took precedence over plants — a style which the 19th century reacted strongly against.[8]

In fact, this was not the first rock garden in the country, and Banks's role in its creation was less central than the legend suggests; on the other hand, the rock garden probably had a far greater horticultural legitimacy than the legend implies.

William Forsyth, who was later to be a co-founder with Banks of the Horticultural Society, replaced Philip Miller as Curator of the Chelsea Physic Garden in February 1771, and began "a great interchange of exotic plants... between the Society and various Noblemen, Gentlemen and others". William Curtis, under the direction of Stanesby Alchorne, the Demonstrator of Plants, was probably most directly responsible for the rock garden's creation. It was not an afterthought, a way of putting a surprise gift of materials to use, but a carefully thought-out experiment. In September 1772, Alchorne "purchased about 40 Tons of old Stones from the Tower of London which he laid into the garden, at his own expense, for the purpose of raising an Artificial Rock to cultivate plants which delight in such soil and begged this committee to accept them". Two additional gifts, one of lava, from Banks, and one of flints and chalk, were incorporated into the growing structure. The "rock", as it is called in the Minutes, was completed in 1774.[9]

It has now been established that Banks's lava did not come from Mount Hekla, and was not a collection of geological curiosities, but had been taken on board his ship to use as ballast for the return journey. Not all the lava was used at Chelsea; he had a further quantity taken up the Thames on barges, and put to use at Kew for an apparently similar rock garden, which has not survived and about which there is little in the way of documentation. In 1784 a French visitor to Kew, Faujas de Saint-Fond, wrote: "since this lava is full of cavities, fissures and rugosities, and moreover is spongy, soaking up water which is retained for long periods, the idea arose of using it for wide, slightly raised, borders surrounding

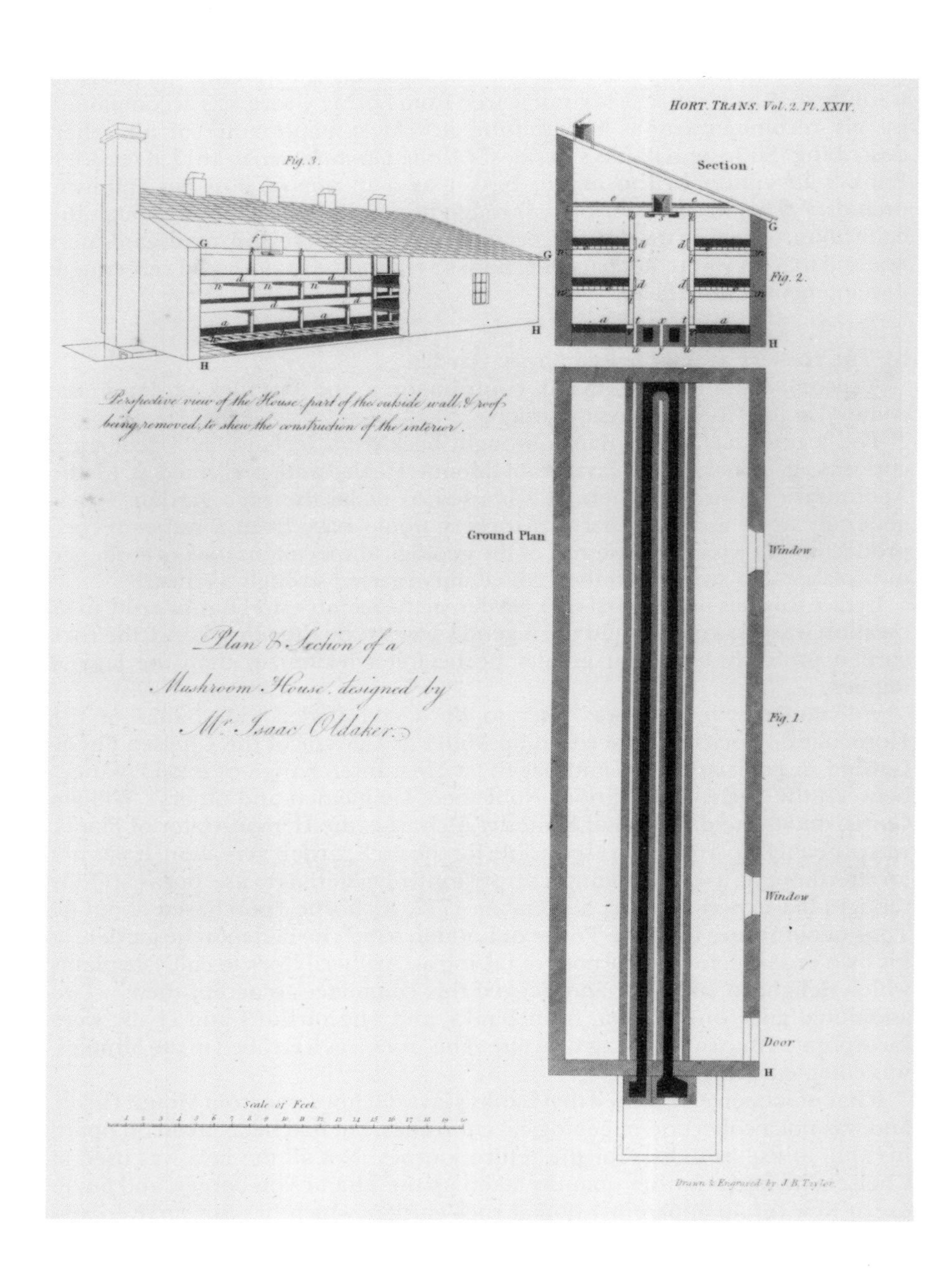
HORT. TRANS. Vol. 2. Pl. XXIV.
Fig. 3.
Section
Fig. 2.
Perspective view of the House part of the outside wall & roof being removed to shew the construction of the interior.
Ground Plan
Window
Plan & Section of a Mushroom House, designed by Mr. Isaac Oldaker.
Fig. 1.
Window
Door
Scale of Feet
Drawn & Engraved by J. B. Taylor.

the beds of a shady moss-garden which would be unique of its kind".[10] The choice of lava for the Chelsea rock garden would appear to have had a sound horticultural motivation.

Nonetheless, this rock garden had no direct progeny that we know of after Banks's time. During the early 19th century, it fell into decline; in 1848, Robert Fortune had a circular pool for aquatics built into its top, and thus made it into something like the structure we know today. The very fact that there is no known illustration of the Chelsea rock garden from its first 70 years of existence does not suggest that it was widely imitated, and the further fact that the place of origin and the functional purpose of the lava were so long misunderstood again suggests that its practical influence on the cultivation of alpines was not great.

c) Banks's garden at Spring Grove

Let us next look at Banks's own garden at Spring Grove in Isleworth, Middlesex, west of London. Banks first leased this property in 1779, and eventually acquired the title in 1808. It was from 1791 that the gardens began to be developed, primarily for use as the family's kitchen garden, but also, and increasingly, to act as an outstation for Kew. It was an eminent garden in its day for its collection of rarities. There are few descriptions of it, but that may be simply a result of the late development of the gardening press in England —Loudon's *Gardener's Magazine* only began in 1826. On the other hand, in view of Banks's reputation for making his plants at Kew inaccessible, it is possible that there were other factors. Certainly after Banks's death the number of plants sent from Spring Grove to the Horticultural Society for exhibition was greatly increased.

Banks's original gardener was John Smith, whom he referred to as "my master", but the most famous of the Spring Grove staff was Isaac Oldacre. Oldacre had worked in St Petersburg as a gardener to Czar Alexander I, and Banks evidently cultivated him for his knowledge of gardening in northern latitudes. He must already have been working for Banks by 1813, for the following year the Horticultural Society sent a deputation to see the mushroom house he had designed on Russian principles. A plan of the building was published in the Society's *Transactions*, with the Society's endorsement.[11] Imitations soon began to be built, and half a century later William Robinson claimed it as the beginning of successful mushroom growing in this country.[12]

As the working garden for a country, or at least suburban, estate, the Spring Grove garden was primarily concerned — in its earlier years — with food production. In this context special attention must be drawn to Banks's experiments with cranberries. He did not introduce these, but he seems to have been the first person in England to grow them for their fruit, and in 1808 he wrote an article for the Horticultural Society's *Transactions* on their culture, saying that they, *Vaccinium macrocarpum*[*on*], had "now become an object of some importance in the economy of the family".[13] It is important to appreciate what a pioneering venture this was on Banks's part, for according to Sturtevant it was not until the 1840s that cranberry culture was carried out on a commercial scale in its native America.[14]

The fame of Spring Grove as a garden did not long survive Banks's time. Lady Banks was succeeded by Henry Pownall, whose gardener Hutchinson greatly modified the gardens. Loudon visited the estate in 1833 and reported on the changes. Oldacre's grape pit had been converted "into the head gardener's house. This is one of the most economical and ingenious modes of procuring a gardener's house that we ever heard of; but we canot say much in favour either of its commodiousness or its comfort."[15] Towards the end of the century the house was replaced, and the gardens remodelled by the firm of James Veitch and Son.[16]

Banks's own practical contributions to gardening were worthy and significant but not such as to set him on a level above his contemporaries. Let us now turn our attention to his role as a horticultural patron — and that means, primarily, to his role in forming the Horticultural Society.

The foundation of the Horticultural Society

Banks's attempts to gather information on horticultural practice long predated the creation of the Horticultural Society. In 1792, for example, he requested that Sir George Staunton should report on Chinese methods of dwarfing trees and forcing plants out of season, on waterlily cultivation, and on the possible introduction of chrysanthemums and tree peonies, as well as fruits. (Staunton's expedition did not produce the desired results).[17]

The Horticultural Society itself was not originally the brainchild of Banks. On 29 June 1801, John Wedgwood, of the famous pottery family, wrote to William Forsyth, the Royal gardener at Kensington, that he was "turning my attention to the formation of a Horticultural Society... If you should see Sir Joseph Banks, will you be so good as to ask him his opinion of the plan, and learn how far we might have a chance of having his patronage of the scheme". Forsyth forwarded this enquiry and on 31 July, Banks replied that he "approve[d] very much the idea". Wedgwood continued to canvas support for his project, elaborating further in a letter to Forsyth the next year, "the Bath Agricultural Society... has given a premium for raising new sorts of apples from the pippin. These appear to be the only instances where any branch of gardening has been encouraged by the agricultural societies, and then only so far as they are considered in an agricultural point of view. It is now proposed to form a society for the sole purpose of encouraging Horticulture in its different branches, to form a repository for all the knowledge which can be collected on this subject and give a stimulus to the exertions of individuals for its farther improvement".[18]

Forsyth was at this time busy preparing his treatise on fruit trees, and it was not until 1804 that action took place. On 7 March of that year, the founding meeting was convened at Hatchard's bookshop on Piccadilly, consisting of Wedgwood, Banks, Forsyth and four others, all associated with Banks: William Townsend Aiton, head gardener at Kew, the nurseryman James Dickson, the amateur botanist Richard Anthony Salisbury, and Banks's old friend and travelling companion Charles Greville. Banks immediately began manipulating the formation of the Society according to his wishes. A couple of weeks later, he wrote to Thomas Andrew Knight, saying "I have taken the liberty of naming you an original member"; Knight replied, giving his assent, "provided the members of it possessed a sufficient stock of scepticism to guard them from imposition".

Knight was effectively Banks's protege. They made contact in the early 1790s, when Banks needed someone in Herefordshire to conduct inquiries for the Board of Agriculture. He became interested in Knight's amateur experiments in plant physiology, and encouraged him to carry his work further. The first fruit of this encouragement was Knight's paper on the inheritance of decay in fruit trees, read at Bank's urging to the Royal Society in 1795. Knight, in these early years, was continually deferential and self-abasing. A typical phrase in a letter to Banks runs: "Your having been so kind as to pay attention to my unimportant Experiments encourages me to take the Liberty of mentioning the Result of some made during the last summer..."[19] Banks found himself continually offering Knight a counselling service: "When you consider your experiments upon the fecundation of plants, and improving the kinds of them by coupling the best males and females of each sort, as unimportant matters, you really act very differently from what I feel myself disposed to do on the ocasion. I am loth to speak in a dictatorial style, if my opinion differs from yours; but I do confess, I think no experiments promise more public utility than those for improving the breeds of vegetables".[20]

Knight had not been involved in the negotiations for establishing the Horticultural Society, probably because William Forsyth was involved. Forsyth had written a pamphlet, the nucleus of his later treatise on fruit trees, vaunting the merits of a "plaister" he had devised for treating tree wounds, and Knight had attacked these claims. There seems no reason to doubt that Forsyth had a degree of success in renovating trees, but such success as he had was due less to the merits of his plaister than to his attendant remedial operations. Some of Forsyth's friends jumped into the controversy ill-advisedly, and issued statements of unconditional support without having examined the evidence; on the other hand, Knight, as George W. Johnson later put it, "poured forth insinuations and charges in a wrathful tone, very unbecoming either a philosopher or a gentleman". He attacked Forsyth's honesty and that of his supporters, made public bets on the lifespan of Forsyth's trees, and accused him of plagiarism.[21] All during the controversy, Banks had managed to stay on good terms with Forsyth while encouraging Knight in his experiments. But when it is remembered that Forsyth was one of the original projectors of the Horticultural Society, and that his supporters were among its most prominent early members, it is quite remarkable that Banks was able to inject Knight into the proceedings at such an early stage.

Forsyth died within a few months of the Society's first meeting; the way was then clear for Banks to push Knight into greater prominence. A meeting was convened in March 1805 to draw up the Society's terms of reference, and Banks, in the chair, had the drawing up of the prospectus assigned to Knight. As a result, the Society, which might have taken any form whatever, was quickly dedicated to encouraging the sort of research that Banks and Knight had been pursuing.

Banks's day-to-day role in the Society quickly subsided, because of gout as much as anything else. He slipped into the congenial role of Vice-President, which did not require a presence at meetings, but entailed consultation on important issues. However, he was roused into a flurry of literary activity, contributing to the Society's *Transactions* 11 articles, mostly within a period of 6 years, including a miscellaneous summary of horticultural observations from French authors. (This number is a mere trickle compared with Knight's 149 articles, but when you recall that Knight's appeared over a 33-year period, his rate of production was only slightly more than double Banks's rate). Several additional articles took the form of letters to Banks, which he passed on for publication.

One of Wedgwood's main points, while canvassing for his intended society, was the lack of awards made to advance horticulture, whereas agricultural societies regularly awarded premiums to cultivators. Part of the concept of the new Horticultural Society was that it should promote horticulture by giving awards.

In 1808 a Large Gold Medal was commissioned, with instructions that "when ready [it] should be sent to Sir Joseph Banks". The design showed a greenhouse in a garden setting, interestingly similar in design to Banks's pineapple house at Spring Grove. The first specimen of this medal was presented to Banks in 1811 "for his unremitted and important services to the Society, from its first institution". The next Gold Medal was presented to Knight, in 1814.

Before long, Council worried that the Large Medal might lose esteem if it was awarded too often. In 1820, on the death of Banks, Council decided to commemorate him with a smaller silver medal, to be known as the Banksian Medal: "the Banksian will be exclusively confined to rewarding the exhibitors of objects transmitted or brought to the general meetings of the Society". In its first year (1820–21), 44 Banksian Medals were awarded. The Banksian Medal, in its various forms, remained the mainstay of the Society's awards to exhibitors until very recent times, and is still used for awards made by affiliated societies.

Thus began the Royal Horticultural Society's long history as a bestower of awards. The Banksian was soon followed, in 1837, by the Knightian Medal, still awarded for exhibits of vegetables. In all, the Society today has 45 different medals, cups and prizes awarded for exhibits at its various shows.[22]

Banks and the promotion of horticulture

Let us now look at the sort of horticultural work that Banks was encouraging, through his own writings and those of his protege Knight.

Banks's first paper in *Transactions*, read in May 1805, was "An attempt to ascertain the time when the potatoe (*Solanum tuberosum*) was first introduced into the United Kingdom". It is an amateur piece of work by today's standards, and its conclusions have not been sustained by Salaman, but it was a pioneering attempt to establish an uncertain date of introduction.[23] Aiton's *Hortus Kewensis* in 1789 had begun the process of attributing dates to plants, on the basis of the Kew records and a variety of published sources. But Aiton's abbreviated style did not allow the leisure to assess circumstantial evidence of the sort Banks examined. In other papers, Banks used Tusser, Parkinson and other gardening writers to date introductions. But he went further than this, and in his paper "On the forcing-houses of the Romans", he drew on the early glimmerings of British archaeology, citing Daniel Lyson's excavations in Gloucestershire for evidence of Roman heating methods.[24]

Two years after Banks's death, John Claudius Loudon published his *Encyclopaedia of Gardening*, which included a massive treatment of the history of gardening. With this work, garden history reached a level of achievement which was not to be rivalled until well within the present century. In the preface he acknowledged his debt to the Banks herbarium, and to the unpublished, and now lost, manuscript history of gardening by William Forsyth Junior — in view of the relations between Banks and Forsyth senior, another possible channel of Banksian influence.[25]

Banks's last paper in the *Transactions* was a further exercise in horticultural historiography: an attempt to trace the history of the arrival and spread of *Aphis lanigera* in England.[26] With this paper, he effectively initiated modern plant epidemiology. Even more importantly for plant pathology, in 1805 he published a separate paper on fungal infections in wheat. In this he drew attention to the frequency with which diseased wheat was found in the vicinity of barberries, which also suffered from a mildew, and asked, "Is it not more than possible that the parasitic fungus of the barberry and that of wheat are one and the same species, and that the seed is transferred from the barberry to the corn?" Unfortunately, neither Banks nor any of his associates attempted to put the matter to an experimental test; although this failure might have been due to the ridicule which the paper received in the agricultural press. At any rate, it was not until 1865–66 that Anton de Bary verified Banks's hypothesis, by reciprocally infecting wheat and barberries with each other's fungus. But although his influence was delayed, through this paper Banks stands as a precursor of the development of modern plant pathology.[27]

Plant pathology was one of the four major developments in 19th-century horticulture I mentioned at the start of this paper; the others were glasshouses, acclimatisation and plant breeding. I will now examine Banks's contribution in each of these areas.

a) Glasshouses

In the Horticultural Society's Prospectus, Knight wrote that: "The construction of Forcing houses appears to be generally very defective, and two are rarely constructed alike, though intended for the same purposes; probably not a single building of this kind has yet been erected, in which the greatest possible quantity of space has been obtained, and of light and heat admitted, proportionate to the capital expended".[28]

In 1809, Banks endorsed these hopes in a paper in *Transactions*: "The public have still much to learn on the subject of *Hot-houses*, of course the Horticultural Society have much to teach..." "The next generation will no doubt erect *Hot-houses* of much larger dimensions than those to which we have hitherto confined ourselves, such as are capable of raising trees of considerable size, they will also instead of heating them with flues, such as we use, and which waste in the walls that conceal them, more than half of the warmth they receive from the fires that heat them, use naked tubes of metal filled with *steam* instead of *smoke*...".[29]

In his pinery at Spring Grove, designed by Aiton and built by George Tod, the flue for heating the pit was "above ground all round" — an early experiment in the sort of heating that Banks anticipated.[30] This was still a fire-heated flue, but experiments in steam heating became progressively more frequent during the first two decades of the century, eventually in turn to be succeeded by heating by hot-water pipes.

Knight and others filled the early volumes of *Transactions* with discussions of greenhouse heating, and the proper pitch of glass roofs, Knight alone contributing 13 papers on greenhouse construction and management; but the real breakthrough came in 1815, with Sir George Mackenzie's paper recommending a spherical roof, parallel to the dome of the heavens, as the means of obtaining optimum lighting for plants at every time of day.[31] The following year, J.C. Loudon patented his wrought-iron glazing bar, and made

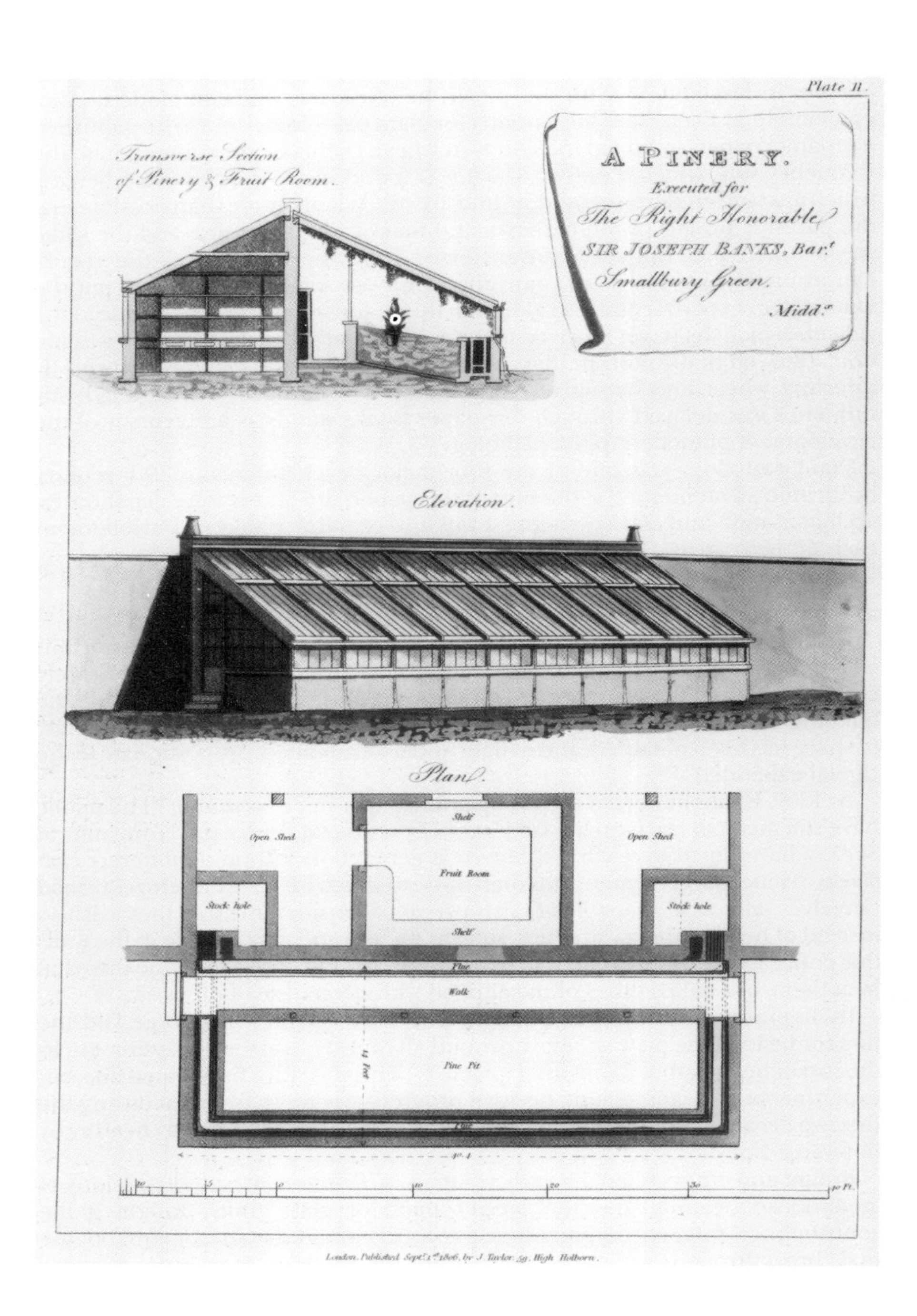

Plate II.
A PINERY.
Executed for
The Right Honorable
SIR JOSEPH BANKS, Bar.t
Smallbury Green,
Midd.x
Transverse Section
of Pinery & Fruit Room.
Elevation.
Plan.
Open Shed
Shelf
Fruit Room
Open Shed
Stock hole
Shelf
Stock hole
Flue
Walk
Pine Pit
Flue
London, Published Sept.r 1.st 1806, by J. Taylor, 59, High Holborn.

curvilinear construction possible. In the wake of Loudon, the 1830s and 1840s were to see Banks's prediction fulfilled, with the erection of the great glasshouses of Chatsworth and Kew.

b) Acclimatisation

Early attempts to introduce exotic plants into English gardens had often failed because of inadequate knowledge of their native growing conditions.

In 1805, Banks published a major paper on "inuring tender plants to our climate". In this he wrote that "several valuable shrubs, that used to be kept in our stoves, are now to be seen in the open garden", and while he acknowledged the possibility that such plants had actually come from colder climates than the circumstances of their introduction suggested, he was convinced that over successive generations, plants could adapt themselves to differing climatic conditions.(32) His experiments on the Canadian wild rice, *Zizania aquatica*, at Spring Grove, seemed to confirm this.(33) "If we could make the Myrtle bear the climate of Middlesex", he hoped, "as well as it does that of Devonshire, or exempt our laurel hedges from the danger of being cut down by severe frosts, it would be an acquisition of no small consequence to the pleasure of the gentleman, as well as to the profit of the gardener.

"Old as I am, I certainly intend this year to commence experiments on the Myrtle and the Laurel".(34)

Whatever the result of these experiments, his hope was not dampened, and in 1809 he wrote, "It does not require the gift of prophecy to foretell, that ere long the *Aki* and the *Avocado pear* of the West Indies, the *Flat Peach*, the *Mandarine Orange*, and the *Litchi* of China, the *Mango*, the *Mangostan* and the *Durion* of the East Indies, and possibly other valuable fruits, will be frequent at the tables of opulent persons; and some of them perhaps in less than half a century, be offered for sale on every market day at Covent Garden".(35) This ideal continued to animate gardeners into the second half of the century. In 1852, Donald Beaton could ask his readers, "Only think of a cottage gardener having a plot of Pine-apples growing at the end of his Rhubarb bed like so many globe Artichokes!".(36)

c) Plant breeding

The possibility of cross-breeding plants had been demonstrated by Thomas Fairchild in the 1720s. By the end of the century, it was beginning to dawn on the gardening world that some well-known economic plants might have a hybrid origin; Knight acknowledged in the Prospectus the lack of information on the origins of many cultivated plants, and Banks's first paper in *Transactions*, seeking information on the wild state of wheat, referred to this knowledge as among "of the greatest desiderata of cultivators".(37)

In the Society's prospectus, Knight laid down a challenge for gardeners: "almost every ameliorated variety of fruit appears to have been the offspring of accident, or of culture applied to other purposes. We may therefore infer, with little danger of error, that an ample and unexplored field for future discovery and improvement lies before us, in which nature does not appear to have formed any limits to the success of our labours, if properly applied".(38)

Banks had long encouraged Knight in attempts at plant breeding, writing to him in 1798: "Your experiments on apples and grapes must be very tedious, but surely the success of those on annual plants, will induce you to persevere".[39] Knight's first paper before the Royal Society dealt with the importance of breeding.[40] Knight was in fact reasoning from incorrect premises; he was convinced, from reports of the increasing debilitude of various famous fruit varieties, that varieties, like individuals, had a limited life span; it was therefore necessary to devise new varieties to replace those which were destined to disappear. This idea was not completely new — Alexander Hunter in his *Georgical Essays* had suggested 14 years as the natural life-span of a variety — but Knight's forceful advocacy had a practical effect on the gardening community.

Knight received three Silver Medals, and one Silver Banksian, for various individual fruits he raised. Few of these fruits survive today. His strawberries, especially the Downton and the Elton, famous in their day, are among the ancestors of the modern large-fruited strawberries, but have themselves long since disappeared from cultivation; some of his currants, and at least one of his plums, lasted into the twentieth century; some of his apples are still around, though not in commerce, and it is possible that his Spring Grove Codlin, named for Banks's garden, survives, though there it is uncertain that the Brogdale tree was correctly named. It is his cherries that have survived best, and all three of his Medal-winning varieties (Waterloo, Black Eagle, and Elton Heart) can still be found today.[41] Nonetheless, it was for his example and his advocacy, rather than his individual plants, that he was important. His pronouncements about the necessity to breed new varieties to replace the doomed and moribund ones of his day led to decades of fruit breeding, which gave us most of the domestic varieties in commerce today.

Conclusion: the theory of horticulture

Knight's papers were edited by Lindley and Bentham, and published in 1841: 78 papers, more than half of them written during Banks's life, detailing his experiments on budding and grafting, on the ascent of sap, on fertilisers and irrigation, on forcing and pruning, on glasshouse culture and plant breeding. Even before Banks's death, Knight was regarded internationally as, in Mirbel's words, the man who would "clear up the chaos of vegetable physiology".[42]

In 1840, Lindley published his *Theory of Horticulture*, dedicated to Knight's memory. It was divided into two sections, the first devoted to germination and plant physiology, proceeding, in an order which virtually every subsequent textbook has followed, from the root to the flower and fruit; the second part dealt with various horticultural operations, from basic culture through propagation and pruning, to "the improvement of races".[43]

Alexander Cramb, the celebrated gardener at Tortworth Court, recalled the impact of the work on a young gardener:

"Dr Lindley's *Theory of Horticulture...* raised horticulture almost to an exact science, providing us with a knowledge of the natural action of plant life and its requirements, and so demolishing that huge monster — empiricism".[44] That derogatory reference to empiricism alerts one to the changing intellectual climate of the 19th century. One cannot imagine such a comment being made by Banks, or indeed by the Thomas Andrew Knight who, in a letter to Banks in 1798, said, "Nature is rarely sufficiently uniform in its operations to allow us to draw general Conclusions from particular Premises" — and of whom Donald Beaton said, "no one was every more anxious than he to guard the young idea against building up systems of practice on a baseless, so-called-scientific foundation".(45)

But, although it was Lindley's generation that made horticultural knowledge systematic, and communicated it systematically, their work was based on foundations laid by Banks and his associates.

Notes

(1) D. Beaton, 'Root action — potting', *Cottage Gardener* (1860), **23**, 313.

(2) Letter from John Graefer to Banks, British Museum [hereafter B.M.] Add. MS. 33978, fol. 57.

(3) *Botanical Register* (1815), **1**, tab. 71.

(4) W. Herbert, *Amaryllidaceae*, London, 1837, 247–8.

(5) *Botanical Register* (1817), **3**, tab. 227.

(6) *Botanical Register* (1818), **4**, tab. 346.

(7) *Botanical Register* (1817), **3**, tab. 220.

(8) Cf. e.g. R. Gorer and J.H. Harvey, 'Early rockeries and alpine plants', *Garden History* (1979), **7**(**2**), 69.

(9) The history of the Chelsea Physic garden rockery has been written by Susan Schnare in her thesis, *Sojourns in Nature: the Origins of the British Rock Garden*; I am grateful to her for allowing me to use her work.

(10) G. Meynell and C. Pulvertaft, 'The Hekla lava myth', *Geographical Magazine* (1981), **53**, 435.

(11) I. Oldaker, 'Account of the method of growing mushrooms in houses', *Transactions of the Horticultural Society of London* [hereafter *Transactions*] (1817), **2**, 336–341.

(12) W. Robinson, *Mushroom Culture: its extension and improvement*, London, 1870, 9–12.

(13) J. Banks, 'An account of the method of cultivating the American cranberry, *Vaccinium macrocarpum*, at Spring Grove', *Transactions* (1812), **1**, 75–78.

(14) E.S. Sturtevant, *Sturtevant's notes on edible plants* (ed. U.P. Hedrick), Albany, NY, 1919, 402. For continued occasional efforts to grow cranberries à la Banks later in the century, see R. Hogg, *The Fruit manual*, 5th ed., London, 1884, 317. Interestingly, Loudon says that in the 1830s, Banks's successor was growing cranberries "in dry soil, not peat, bearing abundant crops"; see J.C. Loudon, 'Notes on Gardens and Country Seats visited...', *Gardener's Magazine* (1833), **9**, 649.

(15) J.C. Loudon, ibid.

(16) H.J. Wright, 'Gardens about London: Spring Grove House', *Journal of Horticulture* (3rd series) (1896), **32**, 425–6.

(17) Banks to Staunton, 18 August 1792, quoted in W.T. Stearn, 'Three letters of horticultural interest from Sir Joseph Banks to Sir George Staunton, 1792–1794', *Journal of the Royal Horticultural Society* (1974), **99**, 339–347.

(18) For the correspondence relating to the founding of the Society, see H.R. Fletcher, *The Story of the Royal Horticultural Society 1804–1968*, London, 1969, 19–36.

(19) Letter from Knight to Banks, 18 January 1798, B.M. Add MS 33980, fol. 129; similarly, 14 January 1799, ibid fol. 171: "... if anything occurred in my unimportant Experiments".

(20) Banks to Knight, 1798, quoted in the biographical sketch accompanying T.A. Knight, *Selection from the Physiological and Horticultural Papers*, London, 1841, 14. See also ibid., 28–29, for Banks's letter of 10 April 1800: "Whether any of our predecessors may have been better qualified to investigate the physiology of plants than you are, I shall not decide upon; but that you are eminently qualified for such undertakings I will most readily declare".

(21) See G. Meynell, 'The personal issue underlying T.A. Knight's controversy with William Forsyth', *Journal of the Society for the Bibliography of Natural History* (1979), **9**, 281–7. For Johnson's remark, see *Cottage Gardener* (1852), **7**, 378.

(22) All unattributed quotations in the section discussing medals are taken from the Society's Minutes of Council. The history of the Society's medals has been written by Mr W.L. Tjaden, and will be published shortly in *Archives of Natural History*.

(23) J. Banks, 'An attempt to ascertain the time when the potatoe (*Solanum tuberosum*) was first introduced into the United Kingdom...', *Transactions* (1812) **1**, 8–12.

(24) J. Banks, 'On the forcing-houses of the Romans, with a list of fruits cultivated by them, now in our gardens', *Transactions* (1812), **1**, 147–156.

(25) William Forsyth, Junior, built up perhaps the most extensive historical library of gardening books in his time, and prepared a manuscript on the history of gardening; Loudon made extensive use of this manuscript, and acknowledged his debt to Forsyth in three different publications. In the absence of documentation, it is difficult to make any statement about the relations between Forsyth junior and Banks; Forsyth's manuscript disappeared after his death. But in view of the relations between Banks and Forsyth senior, it seems to me likely that the younger Forsyth must have derived some degree of encouragement in his researches from Banks.

(26) J. Banks, 'Notes relative to the first appearance of the *Aphis Lanigera*, or the apple tree insect, in this country', *Transactions* (1817), **2**, 162–9.

(27) J. Banks, *Short account of the cause of the disease in corn, called by farmers the blight, the mildew, and the rust*, London, 1805.

(28) T.A. Knight, 'Introductory remarks relative to the objects which the Horticultural Society have in view', *Transactions* (1812), **1**, 5.

(29) Banks, op. cit. (24), 150–151.

(30) G. Tod, *Plans, Elevations and Sections of Hot-Houses...*, 1812, 15, plate 11.

(31) G. Mackenzie, 'On the form which the glass of a forcing-house ought to have, in order to receive the greatest possible quantity of rays from the sun', *Transactions* (1817), **2**, 171–7.

(32) J. Banks, 'Some hints respecting the proper mode of inuring tender plants to our climate', *Transactions* (1812) **1**, 21.

(33) Ibid., 22.

(34) Ibid., 24.

(35) Banks, op. cit. (24), 151.

(36) D. Beaton, 'Stove plants in the open air', *Cottage Gardener* (1852), **8**, 305–6.

(37) Banks, op. cit. (23), 12.

(38) Knight, op. cit. (28), 1–2.

(39) Quoted in Knight, op. cit. (20), 14.

(40) Knight, 'Observations on the grafting of trees', printed in ibid., 81–84.

(41) Quoted in ibid., 16–17.

(42) Here follows a list of Knight's medals from the Horticultural Society. Gold Medal: 1814, for grafts, buds, and papers. Large Silver Medals: 1815, for the Black Eagle Cherry; 1817, for the Waterloo Cherry; 1818 for the Elton Cherry. Silver Banksian Medal: 1822, for new pears. (In 1836 he received the first new Large Gold Medal for his researches).

Of Knight's fruits, the *National Apple Register* lists the Spring Grove Codlin as being in the National Fruit Trials collection if true. His plum 'Ickworth Imperatrice' was used for drying well into the present century. His three prize-winning cherries were still known to Grubb, who reported that seven of Knight's cherry trees were still left at Downton in 1926; see N.H. Grubb, *Cherries*, London, 1949, 126–7.

(43) J. Lindley, *Theory of Horticulture*, London, 1840.

(44) A. Cramb, 'British Gardeners: 19', *Gardeners' Chronicle* (1875), **3**, 720.

(45) Knight to Banks, 18 January 1798, B.M. Add MS 33980, fol. 129; Beaton in 'Root action: potting' *Cottage Gardener* (1860), **23**, 313.

From the last-cited passage, let us take some of Beaton's words as valedictory: "Some say, that we shall never be driven to make the best of what we have, until all the plants in the world are found out, and brought together. Then, instead of sending out collectors at enormous cost, we shall lay out our strength on other means, such as hybridising, forcing, or starving plants above or below their natural ways, to procure 'sports', and trying their capacities for different climates; — and all these points must engage the attention of gardeners some day or other."

A. Frost, 'The planting of New South Wales: Sir Joseph Banks and the creation of an Antipodean Europe', in *Sir Joseph Banks: a global perspective* (eds. R.E.R. Banks and others), Royal Botanic Gardens, Kew, 1994, 133–147.

10. THE PLANTING OF NEW SOUTH WALES: SIR JOSEPH BANKS AND THE CREATION OF AN ANTIPODEAN EUROPE

ALAN FROST

La Trobe University, Bundoora, Victoria, Australia.

In the mid-1850s, William Howitt left the Australian continent with the "fullest conviction" that it was

"destined to become one of the greatest and most flourishing countries in the world. God has done his part. He has planted her amid the southern seas in genial latitudes, and in a position calculated to develop all her resources through unlimited commerce. He has given rich lands for the plough and the pasture; mountains and prairies for the flocks and herds; forests and minerals for her arts; a bosom ample enough and rich enough to nourish myriads of inhabitants; and it depends alone on man whether her progress shall be slow or rapid".(1)

Howitt wrote without any understanding of ecological constraints on development, or of those imposed by twentieth-century Depressions, Recessions and the European Community. However, he did have a distinct premonition of Australia's rapidly-developing character as a "Neo-Europe", which has given it a status in migration history and primary production akin to those of the countries of North and South America.

This character had begun to emerge significantly earlier. By the 1820s (and, indeed, for most of the preceding two decades) the Cumberland Plain, that bulbous lowland that extends about twenty miles to the north, south and west of Sydney was a productive farm. Cattle and sheep grazed in their thousands over its native and introduced pastures, goats and pigs foraged among its bush and orchards, poultry pecked their way through its fields and gardens. Though its soils and climate were not intrinsically well-suited to cereal cultivation, its fields nonetheless regularly offered the European population a year's supply of wheat and maize, while its horticulture was superabundant. James Atkinson described how

"The esculent and culinary vegetables and roots of Europe are all grown in great perfection, together with many others that cannot be raised in England without the aid of artificial heat. Fruits are in great abundance and variety, and many of excellent quality; the principal are oranges, lemons, citrons, peaches, nectarines, apricots, figs, grapes, olives, loquats, grenadillas, pears, apples, plums, cherries, quinces, mulberries, raspberries, strawberries and pomegranates, the whole of which arrive at great perfection, especially such of them as are natives of the south of Europe; the trees are invariably grown as standards".[2]

Even in its detail, such a dispassionate account scarcely conveys the exceptional fecundity of the Cumberland Plain. Better for this purpose is Peter Cunningham's contemporaneous description of the Sydney Market:

"It is held on Thursdays, and attended by individuals from the distance of forty miles or more, with the produce of their agricultural industry. During the preceding day, as you journey towards the interior, you will encouter file after file of carts, loaded with wheat, maize, potatoes, pease, carrots, turnips, cabbages, fruit, pigs, calves, poultry, and indeed all sorts of commodities for culinary use, pouring along the road toward Sydney. A considerable number of oxen are usually sold at the market by auction or private contract, and horses also frequently disposed of."[3]

Better still are the comments of the French-Canadian Patriote François-Maurice Lepailleur, who was the Government's guest in New South Wales 1840–42. Living beside the Sydney-Parramatta road, this native of another Neo-Europe watched awestruck the animals and produce passing east:

"It is incredible to see the amount of hay going to market. One sees hay carts which hold up to two and a half tons. The hay is all pressed down and securely tied like a bundle of merchandise. I saw 500 sheep pass by in a flock this morning and I am sure I have seen flocks three times larger passing by for market and what surprises [me] the most is that all are gelded males. We see lots of horned cattle for market. (4 June 1840)

A large number of fat animals are passing by today — cows oxen sheet etc. (13 July 1840)

As usual lots of fat animals are passing by for market and especially sheep. This is the last time I'll speak of animals, of straw etc. for the market because it's always the same routine. (17 September 1840)."[4]

And these were circumstances replicated in the European settlements that stretched up and down the fertile river valleys that bisect Tasmania.

It was from these two (comparatively) small areas of European settlement that the great mid-nineteenth century pastoral expansion principally went forward. Given direction by Commissioner T.J. Bigge's recommendations that the mainland and Van Diemen's Land settlements should be allowed to develop as civil colonies, and that "the production and growth of fine wool" should be made "the great staple article" of their future exports, [5] this expansion had effectively been set in train a few years earlier by Governor Macquarie's decisions to establish a Government station at Bathurst on the western plains beyond the Blue Mountains, and to grant land outside the Cumberland Plain to explorers. (In 1815, for example, William Cox received 2000 acres at Bathurst, and by 1820 he had 11,000 Merino-cross sheep there. By the mid-1820s, his sons were improving and expanding this flock with Saxon rams, and other pastoralists were availing themselves of the superior bloodlines).

In the 1820s, squatters spread north to the Hunter River valley and New England, west over the Bathurst plains, and southwest into the Southern Highlands and the Murrumbidgee basin, with the Government declaring nineteen counties in 1826. From 1834, Van Diemen's Land graziers brought their swelling flocks across Bass Strait to the superior pastoral district which Mitchell named Australia Felix. By 1850, despite the terrible drought of 1837–9, and the Depression that followed, Merino-cross sheep were to be found all over that broad arc formed by the 800–600 mm isohyets that stretches behind the Great Divide from South Australia up to central Queensland; and the eastern Australian colonies had supplanted Germany as Britain's major supplier of fine wool.[(6)]

By the 1850s, South Australian farmers were exporting wheat to the eastern colonies. Then, in the 1860s and 1870s, as land-granting policies changed and railways probed the interior, selectors in those colonies massively expanded production, particularly in the Murray-Murrumbidgee basins and on the Darling Downs of Queensland. Later, Western Australia followed suit. In these same decades, exchanging herds for flocks, Europeans — now migrating in large numbers — spread into the north of the continent, and from the 1880s, with the development of refrigeration, Australia became a large exporter of frozen meats. Running parallel to these pastoral and agricultural developments was that of mineral exports — first coal, then gold, silver and iron. From the end of the nineteenth century into the 1920s, together with Argentina, Australia stood at the head of what we now call the OECD countries in terms of GDP and *per capita* income. These developments, and its ever growing European population, identified Australia clearly as a Neo-Europe.

Alfred Crosby has pointed out that the Neo-Europes share a number of demographic and geo-physical features. They are "all completely or at least two-thirds in the temperate zones", so that they have "roughly similar climates" — ie, with "warm-to-cool" temperatures and an annual precipitation rate of 150–50 centimetres. They have very extensive temperate grasslands, suitable either for pastoral production or large-scale agriculture. And their indigenous flora and fauna have not provided European rivals with much competition.

It is comparatively easier to understand this last fact where humans are concerned. With their superior social organization, technology and tolerance of micro-organisms (diseases), Europeans have been able comparatively quickly to gain ascendency over indigenous inhabitants, whether in Canada, the United States of America, Uruguay, Argentina, Australia or New Zealand. But, as Crosby asks, "what in heaven's name is the reason that the sun never sets on the empire of the dandelion?"[(7)] It is easier to ask this question than to answer it. Nonetheless, it is abundantly clear that the success of Europeans in the regions they have colonized in the past five hundred years has turned on that of the vegetables, fruits, grains and animals they have taken with them. Crosby cites the example of the Iberians' occupation of the Atlantic Islands, where they "crossed the waters ... with a scaled-down, simplified version of the biota of Western Europe, in this case of the Mediterranean littoral".[(8)] In a similar fashion, the Conquistadores sailed with plants and animals to the West Indies and Central America, for all to spread north and south; while the British, French and Dutch took their foodstuffs and domestic animals to North America. Though each of these groups drew to varying extents on indigenous resources at first, it is safe to say that if their "portmanteau biotas" had not flourished in the new environments, then neither would have their colonizations.

Given the basic similarities, it was perhaps inevitable that Australia (and New Zealand) should have in time matched the countries of North and South

America in character and status. However, what was not inevitable was that it should have done so within the space of one hundred years, for between two and three hundred years of European endeavour underlay the late-nineteenth-century successes of the other Neo-Europes. Inevitably, a number of factors contributed to this comparatively so-rapid progress: the large-scale involvement of central government; superior technology, particularly where shipping was concerned; improved medical knowledge; an efficient labour force. Paramount, however, was Sir Joseph Banks's role in the first decades of the European colonization. It may well seem that in suggesting such a precise beginning point to such a large process I am drawing a very long bow indeed — but let me loose the arrow.

The first thing that had to happen for Australia to develop into a Neo-Europe was that its potential to do so had to be perceived. To say this is only to state the obvious, but consider how easily the opposite perception might have held sway. The first European navigators to see the continent's northern, western and southern coasts were uniform in their dismissal of its potential. Those on the *Arnhem* and *Pera's* voyage to the Gulf of Carpentaria (1623) found "an arid and poor tract without any fruit tree or anything else useful to man". Pelsaert found only "a bare and poor country" with waterless sand dunes in 29°–22° S lat. in 1629. After Tasman's second voyage from the Gulf of Carpentaria to approximately Northwest Cape in 1644, the Batavia Council reported that he had found "nothing profitable, but only poor naked beach-runners, without rice, or any noteworthy fruits". Forty years later, further to the north, Dampier saw "a dry sandy Soil, destitute of Water ... the Woods ... not thick, nor the Trees very big", with inhabitants who differed but "little from Brutes", and who (in doing so) were "the miserablest People in the World".(9) Clearly, the continent and the humans who inhabited it were caught up in a symbiotic relationship of dearth. Even the comparatively much more attractive country about Storm Bay in Tasmania did not engender much enthusiasm in Tasman or the Batavia Council.

Here, it was significant that Banks saw the (again) more attractive eastern coast. As the *Endeavour* proceeded north in late April 1770, he and Cook observed a land "of a moderate height diversified with hills, ridges, planes and Vallies with some few small lawns, but for the most part the whole was cover'd with wood". At Botany Bay, however, beyond the coastal marshes, they found that the woods were "free from under wood of every kind and the trees [were] at such a distance from one another that the whole Country or at least great part of it might be cultivated withtout being oblig'd to cut down a single tree". Away from the marshes, the soil was a "light white sand" that produced "a quant[it]y of good grass", and there were areas of a "much richer ... deep black Soil" which bore "besides timber as fine meadow as ever was seen" (or in Banks's words, "vast quantities of grass"). During their stay of ten days, the pair saw a few people who had no agriculture or domestic animals, no wild animals which might be a food resource, and no "Beasts of Prey".(10)

Now, it is of considerable significance that Banks "saw" this aboriginal landscape through three complementary lenses. First, rather than in the manner of a Dutch trader accustomed to purchasing what others raised, he viewed it with an English farmer's eye. Later, he testified to two House of Commons Committees that, while "the Proportion of rich Soil was small in

Comparison to the barren, [it was] sufficient to support a very large Number of People"; that "he did not doubt but our Oxen and Sheep, if carried there, would thrive and increase"; that this area was fertile enough "to support a Considerable Number of Europeans who would cultivate it in the Ordinary Modes used in England". That is, Banks saw the region about Botany Bay as being capable of supporting the production of the common European foodstuffs.(11)

Banks's second lens acted to confirm and enlarge this perception. In a memorial written about 1717, and published in an English translation in 1744, the Swiss adventurer Jean Pieter Purry defined "climates" by latitudes:

"A Climate is an Extent of Land, contained between two Circles parallel to the Equator, so far distant from one another that there is the Difference of half an Hour in the Duration of the longest Day under each."

To Purry, "climate" meant more than its modern senses, for it comprehended amount and angle of sunshine, temperature and fertility. Numbering that immediately above the equator as the first climate, he pointed to the limited food production within it. Moving north, fecundity improved as fertility did. In the third climate, grain supplanted rice, so that people enjoyed coarse bread. The fourth climate offered "good Bread, good wine, and all Sorts of good Fruit". Being midway between 0° and 66°N latitude, the fifth climate (30°N–36°N latitude) was where daylight attained its median, and was therefore the most salubrious. Within it, the best location was at 33°N, where "we find everything that is good and delicious for the Sustenance of Life" — viz., grains, fruits and "good Wine". Thereafter, Purry saw, the earth's fertility diminished, until above the Arctic Circle, in the twenty-fourth climate, countries were "so dry and barren, that they not only want Bread and Wine, but are not blessed even with a single Tree, and have nothing else but pitiful Brambles". He pointed to North Africa, Syria, Persia and China in support of his paradigm. The fruits of Barbary "are much better and finer than those that grow in *Europe*", he claimed, and it had been Persia that gave peaches to the world.

By analogy, then, since it comprehended the fourth, fifth and sixth climates in the southern hemisphere, New Holland should be fertile. Purry accordingly urged the colonization of Nuits Land: "whence should it be, that all the other Countries of the Earth which are suitable under this Climate, should be good and this alone worth nothing?" He then extended analogy to include minerals: "who knows what there is in *New-Holland*, and whether that Country does not contain richer Mines of Gold and Silver, than, perhaps, *Chili*, *Peru*, or *Mexico*?".(12)

While there is no direct evidence that Banks had read Purry's work, I do not doubt he shared his views. Purry's theory had been widely promulgated by Dr John Campbell's influential compilation of 1744–48; and it informed the proposals for an Antipodean planting put forward by James Matra and Sir George Young in the 1780s, with which Banks was familiar. Indeed, in recommending Botany Bay as the site of first settlement to the 1785 Commons Committee, Banks himself said that he had "no doubt that the Soil of many Parts of the Eastern Coast of New South Wales between the Latitudes of 30 & 40 is sufficiently fertile ..." — that is, he nominated that part of eastern Australia which fell within Purry's fifth and sixth climates.(13)

Still, while it comprehends the just-cited countries, the northern hemisphere latitudinal band of 30°–40° also includes a good deal of water; and the productive European countries lie above it. While, as Purry argued, some of its regions were certainly fertile, their life-sustaining biotas were not ones with which Europeans were familiar, or, indeed, comfortable. Had this band constituted the limits of Banks's conceptualization of New South Wales, he would have equipped the first European colonists in ways very strange to them — e.g., with donkeys and water buffalo rather than horses and oxen, with rice rather than wheat, dates rather than plums.

That he did not was due to his application of another lens. Based on the observation of polar ice rising closer to the equator in the southern hemisphere than in the northern, another belief current in the later eighteenth century was that the former was a good deal colder than the latter. De Brosses discussed this surprizing circumstance in detail; and Callander appropriated his views liberally: "Experience shows that the cold in the *Antarctic* hemisphere is much greater than in ours, and that the seas are found to be frozen in latitudes abundantly temperate in *Europe*". Banks clearly shared this belief, for he told the 1779 Commons Committee that "the Southern Hemisphere [is] colder than the Northern, in such Proportion, that any given Climate in the Southern answered to one in the Northern about Ten degrees nearer to the Pole".(14)

Applying a northerly correction of 10°, then, makes the climate of Botany Bay, in 33°S latitude, "similar to that about *Toulouse*, in the South of *France*";(15) and, with a little bending of meridians, it makes the analogous latitudinal band that spans Western Europe from Lisbon/Valencia/Naples in the south to southern England/Brussels/Frankfurt in the north — ie, a band which encompasses the olives of central and the grapes of northern Portugal; the citrus of Spain; the wheat, maize and vines of the Basin of Aquitaine; the vegetables of northern Europe; the stone fruits of Normandy, Devon, Hampshire and Kent; the alpine cereals of the Massif Central; the peaches, plums, and garden produce of Provence and northern Italy. Within this band were also habitats of all the common European domestic animals, including the fine-woolled merino sheep of Spain.(16) As I shall show, this third filter was crucial in determining how Banks equipped the New South Wales colonists in their first decade, and therefore to their success — let me recall to you Atkinson's comment that it was particularly those fruits and vegetables which were "natives of the south of Europe" that "arrive[d] at great perfection" in New South Wales.

Banks began the business of giving the colonists their "portmanteau biota" with the Toulose analogy to the fore:

"in a Climate similar to that of the South of France which Botany Bay probably is ..."

And his lading covered that latitudinal band just-described:

1) Vegetables: All the common European ones, including bean (green and kidney), cabbage, carrot, cauliflower, celery, pea, spinach
2) Herbs: balm, chive, hyssop, marjoram, mint, parsley, sage, sorrell
3) Berries: gooseberry, mulberry, raspberry, strawberry
4) Fruits: Citrus (lemon, lime, pomegranate, orange, shaddock)
   Stone (apricot, apple, cherry, nectarine, peach, pear, plum)

5) Miscellaneous: garlic, leek, onion, shalott, asparagus, beetroot, parsnip, potato, turnip, chestnut, olive, walnut, lettuce (cabbage, silesia, cos)

6) Grains: barley, rye, wheat (winter and spring)

7) Pastures: clover, lucerne, sainfoin

Phillip reinforced this lading during the voyage out to New South Wales, taking on such acclimatized European varieties as lemon, lime and orange at Rio de Janeiro and the Cape of Good Hope, and adding such exotics as banana, cocoa, coffee, guava, ipecacuhana, jalap, pomme rose (*eugenia uniflora*?), sugar cane, tamarind.(17)

As is well-known, the colonists struggled in the first four years. Much of the seed grain spoiled on the voyage out, some of the vegetables grew but weakly. When Phillip reported the early difficulties, the Pitt Administration fitted out the *Guardian* to remedy them. On her, Banks built a plant-cabin (in the manner of that on the *Bounty*), into which he put an even more extensive range of vegetables, grains and fruits. As we may see, the list fully justifies his assertion to Phillip that he sent "the fruits of Europe". When the *Guardian's* voyage miscarried, Banks and the Pitt Administration tried again; and this time (in September 1791) the *Gorgon* brought its cargo of 200 fruit trees safely into Sydney Cove. More followed on the *Reliance* in 1795, and the *Porpoise* in 1798.(18)

The story where domestic animals is concerned is similar. Banks had told the 1779 Commons Committee that European colonists would need to take such animals as cattle, sheep, pigs and poultry with them to New South Wales. Accordingly, at the Cape of Good Hope Phillip loaded 2 bulls, 3 cows, 3 horses, 44 sheep and 32 hogs, "besides goats, and a very large quantity of poultry of every kind"; and the civil and military officers made a "considerable addition" to this total.(19) Numbers of these animals died during the passage, some were killed by lightning soon after landing, or died from eating rank grass, or were stolen. By May 1788 the totals of all remaining livestock were 7 horses, 7 cattle, 29 sheet, 19 goats, 74 pigs, 5 rabbits, 18 turkeys, 29 geese, 35 ducks, 209 fowls.(20) The Pitt Administration ordered Riou to take on animals and birds at the Cape, which he did, but these were lost in the wreck of the *Guardian*. Parker and King loaded 3 bulls and 23 cows, 4 rams and 62 ewes, 1 hog and 10 sows, 8 pairs of rabbits and 10 pairs of pigeons on the *Gorgon*, some of which died on the passage.(21) In 1792, a few animals also arrived from India. In October 1792, just before Phillip left the colony, the Government stock consisted of 5 bulls and 18 cows, 5 stallions and 6 mares, 105 sheep and 43 hogs; and the totals of privately-owned animals would have been considerably more.(22)

A rigorous process of selection is evident, even if only in a fragmentary way, in the colony's first five years. Worgan recorded that

"The Spots of Ground that we have cultivated for Gardens, have brought forth most of the Seeds that we put in soon after our Arrival here, and besides the common culinary Plants, Indigo, Coffee, Ginger, Castor Nut [,] Oranges, Lemons, & Limes, Firs & Oaks, have vegitated from Seed, but whether from any unfriendly, deleterious Quality of the Soil or the Season, nothing seems to flourish vigorously long, but they shoot up suddenly after being put in the Ground, look green & luxuriant for a little Time, blossom early, fructify slowly &

weakly, and ripen before they come to their proper Size. Indeed, many of the Plants wither long ere they arrive at these Periods of Growth. — but then this Circumstance must be considered, they were sown, the very worst Season.

I have, also, enclosed a spot of Ground for a Garden and make the Cultivation of it one of my Amusements. I put Peas, and broad Beans in, soon after I arrived, (February) the Peas podded in 3 Months, the Beans are still (June) in Blossom, and neither of the Plants are above a Foot high, and out of five Rows of the Peas each 3 Feet in length, I shall not get above 20 Pods, however my Soil is rather too sandy, and in some Spots I see Vegetation has a stronger Appearance. If there are any Plants that flourish better than others, it is thought, that these are Yam, Pomkin; — and ye. Turnips are very sweet, but small. I opened one of my Potatoe Beds, & found 6 or 7 at each Root; Indian Corn, and English Wheat, I think promise very fair; But on the whole, it is evident, that from some Cause or other, tho' most of ye Seeds vegetate, the Plants degenerate in their Growth exceedingly".[(23)]

Still we must make some allowances here. Through necessity, the first plantings were made in the autumn, and before the colonists had been able to make any extensive examination of the area so as to locate the better soils. Phillip fixed the first Government Farm at the little bay to the east of Sydney Cove. By September 1788 Henry Dodd, its superintendent, had established gardens; and had as well 6 acres under wheat, 8 acres under barley, and 6 acres under other grains. This became the colony's first nursery. The crew of the *Sirius* was growing vegetables at Garden Island, and that of the *Supply* on a plot near the hospital.[(24)]

Meanwhile, Phillip had continued to explore, and had located good soil at the head of the harbour. In November, he sent Dodd and one hundred convicts to establish a second settlement there; and over the next two years Parramatta became the centre of the colony's agricultural and horticultural production, with another nursery garden forming about another Government House. This progress was furthered after the arrival in September 1791 of David Burton, the gardener sent out by Banks, who guided settlers to likely locations, and advised them on planting procedures. By 1792, it had become clear that, on the warm, moist, coastal lowland, maize did significantly better than wheat, and barley not at all well. On the other hand, Phillip was able to write that they had "vegetables in abundance. At Parramatta they are now served daily to the Convicts"; that from his own garden he had oranges and "many as fine figs as ever I tasted in Spain or Portugal"; and that his 1000 vines had produced three hundredweight of grapes.[(25)] Another year on, and the pomegranate trees in the Government garden at Parramatta were "well loaded with fruit", while "strawberries served as a border for the paths".[(26)]

We lack precise details of which vegetables flourished. However, Cavanilles asserted that the Government Farm at Parramatta contained "all the vegetables of Europe", and mentioned specifically "kidney beans, peas, cabbages, lettuces, endive, melons, watermelons, potatoes, turnips".[(27)] And certainly a bias towards southern Europe is evident where fruits are concerned — citrus, figs, grapes, pomegranates. As Tench elaborated,

"Vines of every sort seem to flourish: melons, cucumbers, and pumpkins, run with unbounded luxuriancy; and I am convinced that the grapes of New South Wales will, in a few years, equal those of any other country. 'That their juice will probably hereafter furnish an indispensible article of luxury at European tables', has already been predicted in the vehemence of speculation. Other fruits are yet in their infancy; but oranges, lemons, and figs ... will, I dare believe, in a few years become plentiful".(28)

A similar process of selection is evident where animals are concerned, too. While horses did not thrive, the cattle, though few, grew sleek — "[they] have the coats of race horses, glossy as sattin and not a rib to be seen", wrote on observer in 1795;(29) and the Cape ewes grew "too fat to breed".(30) Goats and pigs multiplied quickly. Geese and ducks did well, too, but evidently not chickens.(31)

With this selection process running its course, and learning the initial lessons of the environment, the colonists made very rapid progress in the next half-dozen years. As they extended their activities about Parramatta, as they moved onto the rich Hawkesbury floodplains, and as the numbers of their animals increased, they brought the Cumberland Plain into European life.(32) John Macarthur wrote in August 1794,

"The changes that we have undergone since the departure of Governor Phillip are so great & extraordinary that to recite them all might create some suspicion of their truth. From a state of desponding poverty & threatening Famine, that this Settlement should be raised to its present aspect, in so short a time, is scarcely credible. As to myself I have a Farm containing nearly 250 Acres of which upwards of 100 are under cultivation & the greater part of the remainder is cleared of the Timber which grows upon it. Of this year's produce I have sold £400 worth, & I have now remaining in my Granaries upwards of 1800 Bushels of Corn. I have at this moment 20 acres of very fine wheat growing — & 80 Acres prepared for Indian Corn and Potatoes with which it will be planted in less than a month.

My stock consists of a Horse, two Mares, Two Cows, 130 Goats, & upwards of 100 Hogs. Poultry of all kinds I have in the greatest abundance. I have received no Stock from Government, but one Cow, the rest I have either purchased or bred. With the assistance of one Man & half a dozen greyhounds, which I keep, my table is constantly supplied with Wild Ducks or Kangaroos — averaging one week with another, these dogs do not kill less than three hundred pounds weight. In the centre of my Farm I have built a most excellent brick house 68 Feet in front & 18 in breadth. It has no upper story, but consists of four rooms on the ground floor — a large hall, Closets, Cellar &c. — adjoining is a Kitchen, with Servants Apartments, & other necessary offices. The House is surrounded by a Vineyard & Garden of about three acres the former full of Vines & Fruit trees, & the latter abounding with most excellent vegetables".(33)

Macarthur's success was emblematic of the whole. By August 1799, there were 8682 acres under grain on the Cumberland Plain, and "large tracts of garden-ground". In Government and in private ownership there were 138 horses, 709 cattle, 5103 sheep, 2763 goats and 3459 pigs. In 1805, there were 12,700 acres under cereals; and 517 horses, 4325 cattle, 20,617 sheep, 5123 goats and 23,050 pigs. By this time, the colony was effectively self-sufficient in common European foods; and its European population had risen steadily, by transportation and births, from 1000 in 1788 to 3100 at the end of 1792, to 6000 in 1802 to 7000 in 1805.(34)

Among these statistics are four more-than-usually striking successes. Banks embarked 6 specimens of peach on the First Fleet, and more on the *Gorgon* and the *Porpoise*. It is probable that peach trees also reached the colony on the initiative of others — one East India captain told Banks in 1794, for example, "in my last voyage, I was sent with Convicts, in my way to China, to Port Jackson. I felt for our suffering countrymen, ... and took many doz. of the best Fruit Trees, the growth of the Cape, for their use, which I fortunately landed all well".(35) By the time of the Baudin expedition's visit, there may well have been as many as 160,000 specimens growing over the Cumberland Plain.(36) Péron wrote in awe,

"Of all European vegetables, that which has succeeded best in New South Wales is the peach tree; the cause is either the nature of the soil, which is generally sandy and light, or that the state of the climate is highly favorable to its vegetative faculties. We may see whole fields covered with peach trees, and their fruit is so abundant, that great quantities of it are dried: several of the colonists prepare from it an agreeable kind of wine; others distil from this wine a good-tasted spirit; and it is not unusual to see the farmers fatten their hogs with peaches".(37)

Second, sheep began to reach the colony from Bengal in significant numbers in 1793. When mated with Cape rams, these bred prolifically, dropping up to four lambs at a time. Also, they encountered comparatively few natural predators, so that of the c. 6000 sheep in the colony in 1800, only about 300 had been imported. And when mated with Merino-cross rams, their progeny exhibited fine wool rather than hair. Since others — notably Harold Carter — have told in detail the story of Banks's role in the early development of Australia's Merino-cross sheep, I may summarize briefly here. Initially sceptical of the possibilities claimed by Macarthur, Banks nonetheless did encourage him, Samuel Marsden and others in their breeding efforts, with the results that Marsden successfully shipped 4000–5000 pounds of wool to England in 1811, to be joined in this endeavour in 1813 by Macarthur and Alexander Riley. From these beginnings grew Australia's character as a wool-producing country.(38)

Third, better-fed than in Britain, the women colonists conceived more readily and carried more pregnancies to completion. Better-fed, isolated from the usual diseases of infancy, and enjoying a gentler childhood, the colonial-born European children reached adulthood in proportionately greater numbers than their British cousins: The proportion of children to adult females, for example, rose from 1:5 in 1788 to 1:1 in 1799 and in the same period the percentage of children in the total population rose from 4% to 17%; and reached 20% about 1805 — as the missionary W.P. Crook wrote home in 1804, New South Wales was "an extraordinary place for children".(39) And fourth, by about 1805 this European population had attained a *per capita* income at least as high as the British one — a remarkable achievement after only 17 years, given the constraints of distance, initially unfamiliar environment, and those imposed by the lack of any adequate means of exchange and the Navigation Acts.(40)

In looking at the history of the British colony in New South Wales in its first seventeen years, then, we see a learning/selection phase that lasted about four years (1788–1791); a phase of consolidation and then vigorous growth (1792–1796); and a phase of extremely rapid growth (1797–1805). The encompassing result was the emergence of an Antipodean Europe. The officers of the Baudin expedition who saw it in 1802 were "completely astonished at the flourishing state ... [of] this singular, and distant establishment". There was the township of

Sydney itself, with its substantial government and private buildings; its stone clock tower and stone windmills; its hospital, prison and school; its arsenal, magazines and storehouses. One of these contained "all the articles necessary for the various purposes of domestic life, such as earthenware, household furniture, culinary utensils, instruments of agriculture, &c." and in "astonishing" numbers. In another were "vast quantities of sail-cloth and cordage". There were the dockyards, where the colonists built vessels "from fifty to three hundred tons burthen, ... entirely with the native wood; even their masts are obtained from the forests of the colony". There were the "state ovens", capable of producing 1800 pounds of ship's biscuit per day; and there were the batteries that defended "in the most effectual manner, the approach to the harbour and the town". And this spectacular harbour — as Governor Phillip had said truly, capable of accommodating one thousand ships of the line in the most perfect security — exhibited myriad naval activity. On it, there were "vessels ... from different parts of the world, ... most of which were destined to perform new and difficult voyages". There was the *Investigator*, in which Flinders was preparing to resume his surveying circumnavigation. There were small, locally-built ships engaged in Bass Strait sealing and in the Polynesian pork trade. There were southern whalers, sojourning before pushing east into the Pacific. There were merchantmen bound for China. There were freighters loading with coal for India and the Cape of Good Hope; and there were privateers "on the point of sailing for the western coast of America".(41)

Impressive as these material and economic achievements were, it was the colony's horticultural and pastoral progress that most impressed the French. Beside Government House, Sydney, was a "fine" garden where

"the Norfolk Island pine, the superb Columbia, [grows] by the side of the bamboo of Asia: farther on is the Portugal Orange, and Canary fig, ripening beneath the shade of the French apple-tree: the cherry, peach, pear and apricot, are interspersed amongst the Banksia, Metrosideros, Correa, Melaleuca, Casuarina, Eucalyptus, and a great number of other indigenous trees".

Behind Lieutenant-Governor Paterson's house was a "vast garden", with a "great number of useful vegetables ... in it, ... which have been procured from every part of the world". On the road to Parramatta were pastures of "a very fine and sweet-scented grass, that forms a beautiful verdant carpet, and affords pasturage to numerous flocks of excellent sheep". Interspersed were "spots which have been cleared by the settlers", many of them displaying "pretty habitations" and pastures of introduced grasses on which grazed the "most useful" of Europe's animals. About Parramatta were the striking estates of John Macarthur, William Paterson, John Palmer, D'Arcy Wentworth, William Cox and Samuel Marsden, with their houses and farm buildings, their orchards and pastures, their scores of horses and cattle, their hundreds, even thousands of sheep. Samuel Marsden's Mamre Farm, for example, located seven miles from Parramatta, comprised some 650 acres (103 under cultivations), 10 horses, 26 cattle, 10 goats, 30 pigs and 800 sheep. Sited on a slight hill, the buildings were "spacious and well constructed". The garden was "already enriched with the greater number of the fruit-trees of Europe".(42) Péron wrote in admiration:

"no longer ago than 1794, the whole of this spot was covered with immense and useless forests of Eucalyptus ... This residence ... is ... isolated, in a manner, in the midst of woods; and it was over a very excellent road, in a very elegant chaise, that Mr Marsden drove me to it. What pains, what exertions must have been taken to open such communications! — and these communications, these pastures, these fields, these harvests, these orchards, these flocks, are the work of eight years."[(43)]

As this last quotation indicates, these French officers were so surprized by the New South Wales colony because they had come expecting to find a sparse periphery of Europe — a straggling settlement of ramshackle buildings, uncertain agriculture and scrawny animals, moving only by fits and starts towards viability — and they found instead a flourishing centre. As Baudin remarked, they "could not regard without admiration the immense work that the English have done during the twelve years they have been established at Pork Jackson ... It is ... difficult to conceive how they have so speedily attained to the state of splendour and comfort in which they now find themselves".[(44)]

It is my contention that the fundamental cause of this was the way in which Banks assessed the Botany Bay region's potential to become a Neo-Europe, and that in which he equipped the first colonists to realize this potential. Banks played a central role in each of the three phases of the colony's first seventeen years of existence, and he understood fully the import of what he — and the colonists — were about. In 1797, when reports from the colony were at last vindicating his unwavering views, he wrote to Hunter: "The climate and soil are in my own opinion superior to most which have yet been settled by Europeans". He continued:

"You have a prospect before you of no small interest. To the feeling mind a colony just emerging from the miseries to which new colonists are uniformly subjected, to your abilities it is left to model the rising State into a happy nation. Here matters are different. We have of late seen too many symptoms of declining prosperity not to feel an anxious wish for better times. I keep up my spirits and those of my family as well as I am able, but in truth, my dear sir, could it be done by Fortunatus's wishing cup, I have no doubt that I should this day remove myself and family to your quarters and ask for a grant of land on the banks of the Hawkesbury".

Then, knowing from the discovery of the progeny of the cattle that had strayed in 1788, that European animals might thrive in New South Wales, he added: "I see the future prospect of empires and dominions which now cannot be disappointed. Who knows but that England may revive in New South Wales when it has sunk in Europe".[(45)]

As the quotations at the head of this paper indicate, the processes of Europeanization at work on the Cumberland Plain between 1788 and 1805 continued with ever-increasing effect into the 1820s and 1830s. With help from this region, they were also reiterated in Tasmania from 1803/4 into the 1830s. Then, beginning the great expansion that was to render Australia into a Neo-Europe within fifty years, pastoralists moved north, west and south from the Cumberland Plain, and north over Bass Strait from Tasmania, taking their life-sustaining biota with them. This biota was also crucial to the survival of the South Australian colony in its first years. That is, the biota established with Banks's help on the Cumberland Plain behind Sydney was one of the central factors in the pastoral expansion. This is not to suggest that Banks was the "onlie begetter" of European Australia. The colonists of Tasmania brought some species directly from England (and Rio de Janeiro and Cape Town), as did the West and South Australian ones. Many of the species — appropriately, tropical and subtropical varieties from South America, Africa, India, the East Indies and the Pacific — that underwrote the occupation of northern Australia came from a nursery established at Rockhampton in the 1860s by Antheleme Thozet, a migrant from Mauritius. Still, in a fundamental way, the potential of Australia to be a Neo-Europe was first realized on the Cumberland Plain in the decades about the turn of the nineteenth century, under the guidance of Sir Joseph Banks — so that we may appropriately see him as the central player in this awesome historical drama.

Notes

(1) W. Howitt, *Land, Labour and Gold; or, Two Years in Victoria,* London, 1855, vol. 2, 391.

(2) J. Atkinson, *An Account of the State of Agriculture and Grazing in New South Wales* (ed. B.H. Fletcher), Sydney, 1975 [1826], 57.

(3) P. Cunningham, *Two Years in New South Wales* (ed. D.S. Macmillan), Sydney, 1966 [1827], 37.

(4) F.M. Lepailleur, *Land of a Thousand Sorrows* (ed. and tr. F.M. Greenwood), Carlton, Aust. 1980, 18, 30, 46.

(5) T.J. Bigge, *Report ... on the State of Agriculture and Trade in the Colony of New South Wales,* Adelaide, 1966 [1823], 53.

(6) In 1830 Britain imported approximately 2 million pounds of Australian wool, or 8% of imports. By 1840, this total had risen to 8 million pounds, (22%). By 1850, it had risen again to 39 million pounds (47%).

(7) A.W. Crosby, *Ecological Imperialism: The Biological Expansion of Europe, 900–1900,* New York, 1986, 3–7.

(8) Ibid., 89.

(9) Quoted in G. Schilder, *Australia Unveiled: The Share of the Dutch Navigators in the Discovery of Australia* (trans. O. Richter), Amsterdam, 1976, 93, 116; and in A. Sharp, *The Voyages of Abel Janszoon Tasman,* Oxford, 1968, 316; W. Dampier, *A New Voyage round the World* (ed. A. Gray), New York, 1968 [1927], 312.

(10) J.C. Beaglehole (ed.), *The Journals of Captain James Cook, I: The Voyage of the 'Endeavour', 1768–1771,* Cambridge, 1955, 300, 307–9; J.C. Beaglehole (ed.), *The 'Endeavour' Journal of Joseph Banks,* 2nd edn, Sydney, 1963, vol. 2, 54–60.

(11) Banks's testimony to the 1779 Bunbury Committee is printed in *Journal of the House of Commons* (1778–80), **37**, 311; while that to the 1785 Beauchamp Committee is in Public Record Office [hereafter PRO], HO 7/1.

(12) J.P. Purry, *A Method for determining the Best Climate of the Earth,* London, 1744, 4–32, passim.

(13) J. Banks, 10 May 1785, 'Testimony', PRO, HO 7/1.

(14) C. de Brosses, *Histoire des navigations aux terres australes*, Paris, 1756, vol. 1, 46–76; and J. Callander, *Terra Australis Cognita*, Edinburgh, 1766–68, vol. 1, 30–32; Banks, op. cit. (13).

(15) Banks, ibid.

(16) See H.B. Carter, 'The Historical Geography of the Fine-woolled Sheep', *The Textile Institute and Industry* (1969), January–February, 15–18, 45–8.

(17) For details and sources, see A. Frost, *Sir Joseph Banks and the Transfer of Plants to and from the South Pacific, 1786–1798*, Melbourne, 1993.

(18) See ibid.

(19) See W. Tench, *Sydney's First Four Years* (ed. L.F. Fitzhardinge), Sydney, 1979, 28.

(20) D. Collins, *An Account of the English Colony in New South Wales* (ed. B.H. Fletcher), Sydney, 1975, 22.

(21) Parker, List, PRO, ADM 1/2309.

(22) Collins, op. cit. (20), 210.

(23) G. Worgan, *Journal of a First Fleet Surgeon*, Sydney, 1978, 12–13.

(24) Phillip to Sydney, 24 and 28 September 1788, State Library of New South Wales [hereafter SLNSW], Dixson MS Q 162, 14; and *Historical Records of New South Wales* [hereafter HRNSW], vol. 1, ii, 189–90.

(25) Phillip to Banks, 26 July 1790, 17 November 1791, 2 April 1792, SLNSW, Mitchell MS C213, 58–9, 81–2, 87–8; and 3 December 1791, SLNSW, Mitchell MS A83, 40.

(26) A.J. Cavanilles, 'Observations on the Soil, Natives and Plants of Port Jackson and Botany Bay', in *The Secret History of the Convict Colony* (ed. and tr. R.J. King), Sydney, 1990, 155.

(27) Ibid.

(28) Tench, op. cit. (19), 264.

(29) Palmer to [?], 16 September 1795, Manchester College (Oxford), Shepherd Papers, vol. 10.

(30) Phillip to Dundas, 19 March 1792, HRNSW, vol. 1, ii, 597.

(31) Phillip to Sydney, 28 September 1788, ibid., 192; Cavanilles, loc. cit. (26), 136–7.

(32) My assumption that, once they had manure in abundance, the colonists made use of it perhaps needs some qualification. Caley pointed to unsatisfactory manuring practices in 1803; and the evidence of both Oxley and Best to Bigge c. 1820 suggests that manuring was not then general on the Cumberland Plain. Later Atkinson also complained that the poorest settlers made no attempt to collect and spread manure; and there is similar comment from Tasmania in the 1820s. (See, respectively, G. Caley, 'A short

account relative to the proceedings in New South Wales, 1800 to 1803', SLNSW, Mitchell MS, A79–1, 207; J. Ritchie (ed.), *The Evidence to the Bigge Reports*, Melbourne, 1971, vol. 1, 79, 81–2; Atkinson, op. cit. (2), 32; S. Morgan, *Land Settlement in Early Tasmania*, Melbourne, 1992, 69.

My impressions are, first, that these comments pertain more to small holders who either possessed no animals, or who, if they had a few, lacked the capital to build barns and yards, or who did not see the need of such building in a climate which did not require the housing of animals in winter; second, that having the capital to do so, large holders followed the usual English practice of yarding their animals and collecting straw; and third, that the contemporary complaints concerned the treatment of agricultural land rather than gardens. That is, I am unsure that the complaints do in fact lead us to the conclusion that the manuring of gardens was not widespread.

(33) John Macarthur to [?], August 1794, in J.N. Hughes (ed.), *The Journal & Letters of Elizabeth Macarthur, 1789–1798*, Sydney, 1984, 39.

(34) Hunter to Portland, 30 August 1799, HRNSW, vol. 3, 716; B.H. Fletcher, *Landed Enterprise and Penal Society*, Sydney, 1976, 229; C.M.H. Clark (ed.), *Select Documents in Australian History*, Sydney, 1950, vol. 1, 405.

(35) Bond to Banks, 2 August 1794, Natural History Museum, London, DTC 9, 77–8.

(36) A.C. Cortis, in a letter dated 29 January 1987, makes this estimate.

(37) M.F. Péron, *A Voyage of Discovery to the Southern Hemisphere*, London, 1809, 305.

(38) See H.B. Carter, *The Sheep and Wool Correspondence of Sir Joseph Banks*, London, 1979; and J.C. Garran and L. White, *Merinos, Myths and Macarthurs*, Rushcutters Bay, Aust., 1985.

(39) B. Ganderia and F.M. Forster, 'Fecundity in early New South Wales: An Evaluation of Australian and Californian Experience', *Bulletin of the New York Academy of Medicine*, (1974), **50**, 1081–96; and P. Robinson, *The Hatch and Brood of Time: A Study of the first generation of native-born white Australians, 1788–1828*, Melbourne, 1985, where Crook's opinion is cited, p. 23.

(40) See, e.g. E. Jones and G. Raby, 'The Fatal Shortage — Establishing a European economy in New South Wales, 1788–1805', in *Studies from Terra Australis to Australia* (ed. J. Hardy and A. Frost), Canberra, 1989, 153–67.

(41) Péron, op. cit. (37), 271–6.

(42) Ibid., 273–85, 293–5.

(43) This description, which appears in a note in the French edition [Paris, 1807, vol. 1, 402] was omitted from the English translation. It is cited by A.T. Yarwood, *Samuel Marsden*, Carlton, Aust., 1977, 88.

(44) Baudin to Jussieu, 1 September 1803, quoted in E. Scott, *Terre Napoléon*, London, 1910, 196.

(45) Banks to Hunter, 30 March 1797, HRNSW, vol. 3, 202.

W. Stanton, 'Banks and New World Science', in *Sir Joseph Banks: a global perspective* (eds. R.E.R. Banks and others), Royal Botanic Gardens, Kew, 1994, 149–156.

11. BANKS AND NEW WORLD SCIENCE

WILLIAM STANTON

Department of History, University of Pittsburgh, USA

Joseph Banks's career places him in the company of those commanding figures in each generation of scientists who have been designated the "influentials".(1)

These are men who, though their own observations and theories may be of little moment, by general consent assume responsibility for bestowing the imprimatur of science upon the observations and theories of others. They admit newcomers to the scientific community, and they nudge senior men to pasture. Banks's contributions to his favourite science were not themselves of great moment, but he profoundly influenced the course natural history would take.

The magnitude of Banks's influence owed far less to his botanizing than it did to his seizing the advantage of unofficial contacts (in the way our century so despises) to make room for Daniel Carl Solander, himself and "suite" on board the *Endeavour*. By so doing Banks launched the modern age of discovery. Thereafter no national exploring expedition worthy of the name failed to find a place for the naturalist. In none however were the consequences of such significance as those attending the American effort.

That Banks's legacy should have had the impact it did on science in the new republic owed much to the republic's image in the world's eyes and to the novel position of the scientist in its egalitarian society. Americans must need bear the curse of their continent in world opinion. Buffon and his many disciples had condemned the New World's climate as too "new" and so too hot and too humid to support many and vigorous species of plants and animals. The unfortunate climate had rendered the Mohawk a beardless coward and was sissifying the transplanted European.(2) For observers abroad further confirmation of creeping degeneracy came with Americans' choice of the republican form of government. The confidence even of many at home was shaken as, in the 1830s, their compatriots approached the wilder shores of romantic egalitarianism, proclaiming not just the equality of peoples but of Americans one with another.

Refuting the slander early became a patriotic duty and science part of the intellectual baggage of every man of learning: scientific investigation would restore the Mohawk's manhood and scientific accomplishment compel the Old World's respect, perhaps even awe.

Here were two varieties of egalitarian belief: the equality of peoples and the equality of citizens. Though inseparably joined, they would jog along as much in discord as in harmony. In the 1820s President John Quincy Adams, finding the republic in danger of becoming an intellectual pariah among the nations, announced that the time had come to direct a backward flow of knowledge across the Atlantic, in his phrase to "return light for light". To that end he advocated a program for the administration of government that appealed chiefly to the learned classes: federal aid to the arts and sciences, establishment of a national university, an astronomical observatory, and an exploring expedition to the Northwest Coast to follow upon Lewis and Clark's. Immediately denounced as an aristocrat (he was also known to have installed a billiard table in the White House), his prospects for re-election obliterated, Adams would be the last American president of the century who possessed a knowledgeable sympathy with the aspirations of science.[3] As it happened a project was already abroad upon the land, one of impeccable egalitarian provenance, and one which, marinating in the popular mind, in ways both unforeseen at home and unimagined abroad, would ease acceptance of much of Adam's elitist program. Banks and Cook had stamped the seal of science upon Cook's first voyage merely by sailing with it, in the gentlemanly company of Daniel Carl Solander. The admiralty was sending a vessel to foreign parts and, perceiving the opportunity to botanize, Banks had only to bend the right ear. In the egalitarian society, by contrast, launching a scientific expedition was a rather more complex affair.

The American expedition began to take shape in the popular mind with a startling announcement from one John Cleves Symmes. No scientist, but rather an army veteran and sometime Ohio farmer of inquiring mind, Symmes (1780–1829) was just the man to belie the egalitarian assumption that the outer reaches of eccentricity are the exclusive preserve of the over-educated. He was also just the man to excite a citizenry just then exploring the outer reaches of egalitarianism, with proposals for an ambitious national enterprise that would show a skeptical world what their nation could accomplish in science. Reasoning from first principles, such as the sublime economy of nature, and from observed fact, such as the ubiquity of hollowness and concentricity (e.g. bone, feather, egg, iron fillings over magnet) and the annual disappearance of swallows from the roof tops of Cincinnati ("It appears to be unknown from whence they come, and whither they go"), Symmes concluded that there were holes in the poles of the earth (at "twelve or sixteen degrees"), which was hollow and habitable within. To proclaim his "Theory", as he called it, Symmes in the spring of 1818 dispatched his "Circular Number 1" to the world's learned societies and heads of state. Did his Circular perhaps catch the eye of the Royal Society's busy president?

As cultural patriot, Symmes claimed for the egalitarian republic the honor of discovery and exploration of the holes, or the "verges", as he called them. "I ask one hundred brave companions, well equipped ...; I engage we find a warm and rich land, stocked with thrifty vegetables and animals, if not men ...".[4] Lecturing at first to audiences in the western towns, and later in the eastern cities, Symmes found himself acclaimed as the New World Newton, the Franklin of the West. He received the acclaim with an appealing modesty. When asked, like Newton, how he was able to make "further discoveries than others", he freely confessed, like Newton, to standing on the shoulders of giants who had provided "all the numberless prior discoveries, ready formed and stated, as a base whereon to found new ones".[5] (Inevitably, modesty wore thin at times, and ridicule led him sadly to reflect on the difficult path the eglatarian society reserved to genius. Original thinkers above all others required society's fostering care, for they were proverbially "inattentive to pecuniary matters". Were they not, "their minds would not be at liberty to fully investigate the truth, by deep research ... and by bold speculation".).[6]

Symmes' disciples called upon their fellow citizens to "unite their efforts with him, to obtain for our common country an unequalled fame ...".[7] The issue lay between the self-taught, learned, and modest, and the ignorant and assuming, between East and West, between America and Europe, between plain folk concerned for the honor of their country and sophisticates indifferent to such mundane considerations. Memorials began to arrive in Congress in 1822, the flow increasing as Symmes extended his tours to enhance lives in the towns and cities of the East. One bore the signatures of fifty members of the Pennsylvania legislature. American naturalists, jealous for the national reputation in science, withheld endorsement.[8] They also withheld ridicule, for they promptly perceived, holes in the poles or no, that a national exploring expedition held out exciting possibilities for the naturalist in a society committed to equality and bearing the curse of that commitment in world opinion. Joseph Banks might proclaim that science knows no nationality, but evidence abounds that these men sought the means to remove the curse by upgrading American science.

In that attempt naturalists suffered peculiar handicaps. True, the bounty of species and genera yet unknown to science that the continent afforded made them the envy of their European counterparts. But their novel society lacked the patronage that aristocracy afforded to the traditional society. Of course, like Banks, they sought first of all the means of satisfying their curiosity about the hidden parts of nature. But for them, unlike Banks, to live in science, as they were determined to do, meant living by science. In a country that prided itself on the bounty of opportunity afforded its citizens — a bounty that rendered egalitarianism a usable social ethic — the opportunities to live by science were remarkably few. The biographies of early American naturalists abound with tales of youthful escape — from the family farm, tannery, countinghouse, the backwoods lectern, and other respectable livelihoods. The geologist James Dwight Dana fled the family hardware store to assist in the chemical laboratory at Yale. The botanist Asa Gray survived on a pittance as librarian at the New York Lyceum of Natural History. The entomologist Thomas Say made his bed beneath the skeleton of a horse in the Academy of Natural Sciences.

One young memorialist of Boston's Society of Natural History, desperate to devote himself to conchology, came dangerously near lamenting the absence of aristocracy in the Republic. After taking due note of the commercial advantages that might be expected from an exploring expedition — a sop to the ethic of useful work[9] — he set about linking the welfare of the American scientist with the national honor. "The peculiar circumstances connected with the settlement & subjugation to the purposes of life on a new soil have hitherto been the reason that little attention has been given to the Nat. Sciences & few men have had either the opportunity or the means of making much advance in them. But the time has come when great numbers of learned men are turning their attention to these subjects." Unfortunately, the sciences afforded "little if any recompense to their devotees except the gratification they derive in their pursuit". If the scientists of America only had the benefit of the patronage enjoyed by their counterparts in Europe, "America might hope to boast of a Buffon, a Latreille, a Cuvier" — or, he might have added, a Banks instead of a Symmes. He might have ventured further into heresy, had he not perceived in the proposal of the New World Newton's the possibility of government's performing as patron.[10]

Spotting a door half-open, scientists pushed, and the spate of circulars, memoirs, memorials, and petitions began to tell.

The question of whether "scientifics" should sail aboard ships of the navy never arose. That was a settled point for every exploration of consequence since Banks put to sea with Cook. But, because former expeditions had assigned all scientific duties to one or two persons, it was felt that they had been less than genuinely scientific affairs. Moreover, natural history had made such strides since Cook's day that it was no longer possible for one man to master more than a few of its branches.(11) What was wanted, the expedition's chief publicist argued, was a "Scientific Faculty, complete in all the departments", which would facilitate "a perfect system of reciprocity. The botanist, while plucking a flower would not overlook the insect feeding upon it; because the entomologist would repay the courtesy by gathering from him a plant ...".(12) In a word, this would be the very model of a modern scientific exploring expedition.

The republic, in a fit of cultural patriotism, was determined to out-Cook, Cook. Two years of rancorous quarrels amongst cabinet, Congress and navy officers over which and how many vessels should sail, and how many scientists and artists should join them, severely eroded the original plans, but in 1838 six vessels did set sail with five hundred officers and men and a scientific corps of eight. Geoconcentricity and holes in the poles were forgotten, having proved less diverting, perhaps, than government's interesting attempt to act as a patron of science. The vexed question whether the scientists should be naval officers or civilians was rendered irrelevant by the discovery that scientists were not to be found in the navy's ranks. That in turn caused difficulty in finding an officer of rank willing to command a squadron that shipped civilians. To dispel what doubt might conceivably remain about whose expedition this was, officers, scientists and artists were all to be American citizens, and indeed all but the Scots gardener were American-born.

By any measure the expedition proved to be a grand success. In the course of its four-year circumnavigation it surveyed and charted some two hundred islands in the Pacific (its charts of the Fijis were still in use in World War II), and it established the existence of a southern continent. The flood of natural history specimens with which it returned would sweep aside the inherited commitment to minimal government to make room for science. For no one proposed to jettison these trophies as government had done in casting adrift the birdskins, bones, and plants gathered by Lewis and Clark four decades earlier. Suitable accommodation for the collections was provided when another obliging Englishman, James Smithson, posthumously awarded the egalitarians across the Atlantic an institution for the increase and diffusion of knowledge amongst them. The natural history collections, displayed as trophies the American people had won in their first grand venture into science, soon ranked with Niagara Falls as a tourist attraction.

Publication of the narrative of the expedition and its charts and scientific reports would demonstrate that an egalitarian people could, after all, contribute significantly to culture. To that end it was again decided that scientists working up the scientific results were all to be Americans. But that was before it became apparent that science enough was not to be found in the Republic. The rule was relaxed, and an algologist and mycologist(13) were imported, at the behest of the community of botanists, who remained ever aware how much the reputation of

American science and how much their hopes for government patronage rested upon the quality of the reports. Publication of the results in nineteen handsome quarto volumes, designed to show the Old World what the New could produce in books as well as in science, continued sporadically for sixteen years after the Expedition's return. Congress had mandated distribution of these examples of American science to the world's heads of state, all of them recipients, some years before, of John Cleves Symmes' "Circular Number 1". Members of Congress from time to time would get the wind up over the cost of bringing out "very heavy works" on "bugology" and other "unimportant and useless branches of natural history".[(14)] But such mutterings were regarded as so many howls from the backwoods. Most understood, as the Newton of the West had earlier understood, the need to keep the world informed of America's progress in science. But to avoid offending the backwoods, it was decreed that the botanists forgo the use of Latin in their monographs. As most Americans could not read Latin, its appearance would be "Anti-American".[(15)] To avoid amusing colleagues abroad, the botanists, led by Asa Gray and John Torrey and with moral support from Sir William Hooker, resisted and finally got their way, though only after a two-year delay.

Yet the full measure of the Expedition's significance would lie chiefly in the ways it would shape the future of science in the United States. Together with the natural history surveys the states were then conducting, the expedition provided the first opportunity many scientists had of living by their calling and, by that much, conferred liberation from economic want and society's nagging sanction to perform useful work. By creating the first great federal institutions of science (the Coast Survey, authorized in 1807, aside) the expedition extended the opportunity to future generations. Its natural history collections became the National Museum. Its astronomical apparatus and the simultaneous observations conducted at Washington in the course of its cruise constituted the founding of the Naval Observatory.[(16)] (By way of welcome, astonished astronomers at Greenwich, Berlin, Paris and Vienna presented the new institution with a library). Its live plants became the United States Botanic Garden and its dried plants the United States Herbarium.

The new institutions signalled recognition, by a people committed to least government and to the idea that public science was incompatible with egalitarianism, that advancing science might nevertheless be considered government's proper concern. Employed by Britain, France, and Spain, Joseph Banks's example served to animate the modern tradition of scientific exploration. The legacy was to science. When the new American republic took up the example, rather gingerly, the flow of benefit was less exclusively to science, for there, by countering at one stroke both the slander of the New World and the endemic egalitarian contempt for intellect, it worked a revolution.[(17)] Sir Joseph's legacy, harnessed to hemispheric pride and Captain Symmes' engaging lunacy, yielded the American republic the means to transcend the limitations its politics imposed upon its science. It was a development necessarily unimaginable to Banks, and perhaps even to his overseas friend and colleague at the Royal Society, Benjamin Franklin, when they had comfortably agreed, and even acted on, the highminded principle that science knows no nationality.[(18)]

Notes

(1) The concept of the "Influential" is Donald Fleming's: D. Fleming, *William H. Welch and the Rise of Modern Medicine*, Boston, 1954, 131–2.

(2) On the slander of the New World see the magisterial study by A. Gerbi, *The Dispute of the New World: The History of a Polemic, 1750–1900*, Pittsburgh, 1973.

(3) Not all ears were deaf to Adams's plea. Promoters of an astronomical observatory argued that it would provide an American prime meridian, the monarchial Meridian of Greenwich being unsuitable for the republic: "Are we truly independent, or do we appear so," when on leaving his native shores every American is "under the necessity of casting his 'mind's eye' across the Atlantic, and asking of England his relative position"? The practice was calculated "to wean the affections ... What American seaman has not experienced this moral effect?" Nauticus, 'On a First Meridian', *National Intelligencer* (Washington) (1819), February 2.

(4) See *Western Spy* (Cincinnati) (1818), April 24; *Western Spy* (1818), Supplement, November 16. Symmes sensibly appended a certificate of his sanity. J.C. Symmes, 'Cimmerian Theory ... A Card for D.P. ...', *National Intelligencer* (1819), December 17. The popular success of Symmes' theory, offspring of cultural patriotism, owed much as well to the widespread literacy that the egalitarian ethic encouraged. It was perhaps as someone has succinctly observed in another connection, "Increased literacy releases latent lunacy"

(5) J.C. Symmes in *Western Spy* (1819), October 23. Still when an admirer compared him to Columbus, Symmes observed on reflection that his own method of reasoning was the more scientific: Columbus "was led by facts ... to conclude that there was a western continent, and was gratified by discovering it. I must therefore surely be *more justifiable* in my conclusion ... as I was first led to decide on it upon principle alone ..." *National Intelligencer* (1821), December 10.

(6) J.C. Symmes, 'Cimmerian Theory', *Western Spy* (1819), September 11; *National Intelligencer* (1821), August 18.

(7) Scientia, 'Symmes Theory', *Cincinnati Advertiser* (1822), November 19. Symmes's admirer "Franklin" pointed out that if the monarchs of Europe, who traditionally believed it the "most glorious action in their respective reigns to encourage enterprise of discovery", had instead subscribed to the American notion that it was "an unwise policy to interfere with individual enterprise", our "minds would have been compressed, and science been an insignificant name". 'To the Editors', *National Intelligencer* (1821), September 17.

(8) It was said that the "silence of all scientific men on the matter" only indicated that they privately thought Symmes correct, as it was "mortifying to the pride of science that more discoveries have been made by untaught geniuses than by regular savans". Scientia, op. cit. (7). The respected mathematician Thomas J. Matthews, who, though he saw in Symmes a man in whom

"observation becomes the mere pioneer of fancy, collecting a heap of learned rubbish", nonetheless urged a polar expedition to "render us better acquainted with the arctic zone" and perhaps reveal a northwest passage. T.J. Matthews, *A Lecture on Symmes' Theory of Concentric Spheres, Read at the Western Museum,* Cincinnati, 1824.

(9) The Philadelphia geologist J.P. Lesley, himself adept in the art though deploring its necessity, labelled the practice of emphasizing to legislators the utility of this or that scientific endeavour (especially the state geological and natural history surveys of the first half of the century) "manufacturing public sentiment". The Smithsonian Institution's first Secretary, Joseph Henry, called such appeals, "oblations to buncombe".

(10) [Augustus A. Gould?], Draft of a memorial for the Boston Society of Natural History, n.p., n.d., Museum of Science, Boston. The American perhaps misapprehended. Cf. David Brewster's observing with some bitterness in 1830 that there was not a scientist in England who enjoyed a pension, an allowance, or a sinecure "capable of supporting him and his family in the humblest circumstances!" He found "the spirit of science subdued" and "the honours of successful inquiry" merely "paltry". Quoted in D.E. Allen, *The Naturalist in Britain: A Social History,* London, 1976, 86. If an English scientist could legitimately voice that lament, his American counterpart could do so in spades.

(11) W. Stanton, *The Great United States Exploring Expedition of 1838–1842,* San Francisco, 1975, 41.

(12) Ibid., 47–58.

(13) William Henry Harvey and Miles Joseph Berkeley, respectively.

(14) Stanton, op. cit. (11), 356.

(15) Ibid., 336.

(16) And by extension the Harvard Observatory as well. William Cranch Bond, the Cambridge, Massachusetts, watchmaker and astronomer, was commissioned to conduct simultaneous observations at his own observatory. This recognition by government resulted in an invitation, after the Expedition's return, to establish an observatory at Harvard, which then served as the model for the new observatory in Washington. Bond's son and fellow astronomer recalled that what the Expedition conferred on the father was "that material aid, in scientific pursuits, which is so necessary to success" and he might have added, so rarely to be found in the United States. Ibid., 365.

(17) "To say that the Athenians built the Parthenon to worship themselves would be an exaggeration"; observes David Lewis, "but not a great one". With the slander of the New World in mind one might with no greater exaggeration say the same of Americans and their scientific institutions. D.M. Lewis, et al. (eds), *The Cambridge Ancient History*: vol. 5. *The Fifth Century B.C.,* New York, 1992, 139.

(18) See, *inter alia,* H.C. Cameron, *Sir Joseph Banks, K.B., F.R.S., The Autocrat of the Philosophers,* London, 1952, 210–11; H.B. Carter, *Sir Joseph Banks, 1743–*

1820, London, 1988, 376; B. Hindle, *The Pursuit of Science in Revolutionary America, 1735–1789*, Chapel Hill, N.C., 1956, 220–2.

For more on the expedition see D.B. Tyler, *The Wilkes Expedition, the First United States Exploring Expedition* (*1838–1842*), Philadelphia; H.J. Viola and C. Margolis, (eds.), *Magnificent Voyagers: The U.S. Exploring Expedition, 1838–1842*, Washington, 1985.

G. Métailié, 'Sir Joseph Banks — An Asian Policy', in *Sir Joseph Banks: a global perspective* (eds. R.E.R. Banks and others), Royal Botanic Gardens, Kew, 1994, 157–169.

12. SIR JOSEPH BANKS — AN ASIAN POLICY?*

GEORGES MÉTAILIÉ

CNRS – Museum National d'Histoire Naturelle, Paris, France

During the 17th century, several natural products from China — mainly medicinal — had become important economic products, notably rhubarb, already quoted by Marco Polo; China root,[1] known through the Portuguese merchants of the 16th century; more recently, ginseng, and of course, tea, which, before becoming a popular drink in England and France during the 18th century, played a controversial role as a remedy.[2] So when, during the 18th century, the general tendency of European governments was to promote exploration in the various part of the world, China raised a special interest. And this interest was surely highly stimulated by the cautious attitude of the Chinese authorities towards foreigners. The great difficulty of access to the country could not but increase desire and imagination.

On the other hand, with the development of colonies, the discovery of new natural products highly favoured the idea of acclimatization. In this process, botanic gardens played a key role, in Europe and in the European dependencies in other parts of the world: they were centres where new plants were introduced to be tested and multiplied before being transplanted or sown in other places. We must not forget to acknowledge the part played by the embarked gardeners, botanists and also plant collectors sent to various places, whose skill was crucial for the success of this policy.

Beside the economic aspect of naturalistic investigation, the purely scientific side was important too. The great number of plants, and the strangeness of animals discovered, stimulated the progress of classification which, in turn, gave more efficient tools to list, describe and name the elements of the floras and the faunas of newly explored lands. And so, the interest in dried specimens was not less than the one in living plants or animals.

If we look for a character who would epitomise this general situation, we cannot but meet Joseph Banks. As he said: "Being the first man of scientific education who undertook a voyage of discovery and that voyage of discovery being the first which turned out satisfactorily in this enlightened age, I was in some measure the first who gave that turn to such voyages."[3] "Equally important", as Ernst Mayr points out,[4] "was that Banks gave social and scientific prestige to natural history". He was a passionate naturalist and, as would be said today, an outstanding scholar with a great audience. Thanks to his

* I would like to thank Julia Bruce, Gina L. Douglas and Françoise Serre for their kind assistance. Also Laboratoire de Phanérogamie and Bibliothèque Centrale for their kind permission to reproduce figures no. 2 and nos. 1, 3 and 4 respectively.

financial situation, he could create at Soho Square his own research centre with a library "not ill furnished on subject of natural history", as he modestly wrote to Loureiro[5] or, as Cuvier[6] said in 1821, "rich and very convenient to use thanks to the method which had governed its distribution". Correspondent of scientific institutions in Europe since 1772, President of the Royal Society for 40 years, he was the centre of a network which kept him informed of any advance in the scientific field. In return he would generously offer the fruits of his own information and experimentation. And last but not least, he was counsellor to King George III and to the East India Company. If one adds the real power he had on the organisation of Kew Gardens and his influence on the management of other botanic gardens[7] in India particularly, it seem obvious that Banks was in a position to bring into play or, at least, favour a project which he had in mind since his youth, of the exchange of natural products on a world level, to the benefit of populations and trade within the British Empire. To quote Cuvier again: "He has spread over all gardens of Europe the seeds from the Southern Sea as he has distributed ours in the Southern Sea".

In this world scheme, China, though not a British dependency, seemed to be of particular interest for him. In this respect, he was not alone, as we have just mentioned. From archives, in France and Russia at least, we know that during the 18th century, missionaries in China were regularly asked by Academicians in St Petersburg or ministers of the French government about the economic products of that country. The concern for possible plant introductions was obviously present. In the case of tea, for instance, the questions were about botany, culture, preparation, prices.[8] This source of information did not prevent Pierre Broussonet,[9] soon Secretary of the Royal Agricultural Society in Paris, in a letter dated 28 January 1785,[10] from consulting Sir Joseph Banks about a project of tea plantations in Corsica. The quick answer, acknowledged by Broussonet on 16 April[11] of the same year and followed in less than a month by two parcels of tea-plants, is interesting for many reasons: firstly, it shows the knowledge of the subject with which Banks was credited by a respected French naturalist, and second, it shows that besides illuminating the technical or scientific aspect of the question, Banks was also very efficient to facilitate its fulfilment — actually he advanced £50 for the plants and supervised their dispatch.[12] So, if France did not become one of the main tea producers, this cannot be attributed to the lack of goodwill of an eminent English gentleman. The answer given to a similar request three years later, in 1788, from Lord Hawkesbury, President of the Board of Trade, about the possibility of tea culture in East and West Indies dominions, followed by the famous letter of 27 December 1788 to Francis Baring, Deputy-Chairman of the Court of Directors of the East India Company, about the possibility of cultivating Chinese tea in India, had a different significance. For phytogeographic reasons, based upon his researches, Banks concluded that the most suitable place for the setting of experimental culture of Chinese tea plants should be "the Country (...) which lies between Bengal and Boutan where in a few days journey you get from the Tropical heats and consequently Tropical Productions to a climate similar to that of Europe".[13] Actually Robert Kyd, who was the founder of the Botanic Garden of Calcutta (18 May 1787), had tea plants growing in his private gardens since 1780 and was very eager to get seeds or plants from Canton through the supercargoes or the captains of the East India Company. Banks considered that the garden should first receive the tea plants from China and then dispatch them to a more suitable place, after a first adaptation. But before that, he also recommended the importation of skilled Chinese tea growers and producers, the techniques being still not mastered at all, to train the local workmen. Four years later, he assisted in the preparation of Lord Macartney's Embassy to China, and it seems that seeds of cultivated tea plants were brought back from China and planted in the garden in Calcutta "according to his directions".(14) Unfortunately Banks would not see the success of this project but his point of view on tea planting in India was to be followed, not always with the goodwill of the East India Company, after the end of its monopoly on Indian trade. But, eventually, on 24 January 1834, Lord William Bentinck, Governor-General of India, appointed a Tea Committee which was charged to submit to the government "a plan for the accomplishment of the introduction of tea culture into India and for the superintendence of its execution".(15) At last, Chinese tea could be introduced on a very large scale, but real success came with the cultivation also of a local tea variety which had been found growing wild in the Assamese hills and which now has almost completely overshadowed the Chinese tea shrub in India.

The case of tea illustrates the preoccupation to enrich the natural resources of a developing colony, India, with a new plant of particularly high economic value. Banks was also asked by the East India Company to find information on Chinese hemp culture and processing. One could cite also the quest of the bread-fruit from the Pacific for the West Indies. But this "colonial" preoccupation did not prevent Banks from being very concerned not to let colonial economies hold their own against the mother country(16) on one hand or from considering also the possible introduction of plants — or animals(17) — into Great Britain to the benefit of the local agriculture.

The case of the Japanese medlar or loquat(18) is worth quoting here as exemplifying the global aspect of the problem of plant introduction from China for Banks. Thanks to the catalogue made by Jonas Dryander,(19) we have a precise knowledge of the books in the Banksian library and so, of the sources on which Banks could rely.

In *Flora Sinensis*(20) (published in 1656) the Jesuit missionary Michael Boym describes the fruit of a tree from China called *pipa:* he gives a rather crude picture of fruits and leaves (fig. 1). He stresses the beauty of the leaves and flowers and the extreme sweetness of the fruit.

Half a century later, among the 400 plants from China from the Cunningham collection, described by Leonard Plukenet in his *Amaltheum botanicum* (1703–1704),(21) appears the description of *Arbor sinica Pipa* with another rather crude picture (fig. 2).

On the other hand, in the fifth chapter of *Amoenitates exoticae*, the narrative by the German doctor, Engelbert Kaempfer, of his journey to Japan, published in 1712,(22) one finds, among the class of Pomiferae et Nuciferae plants, the brief description of a plant called *bywa* or *kuskube.*

In 1784, Carl Peter Thunberg in *Flora japonica*(23) gives a diagnosis and botanical identification for this plant but no illustration. He names it *Mespilus japonica*, and gives the Japanese names as *biwa* and *kuskube.*(24)

The *Flora cochinchinensis* by João Loureiro, published in 1790 — the manuscript of which Banks read and criticized(25) — mentions a small tree

FIG. 1. The *pipa* fruit and tree, in M. Boym, *Flora Sinensis*, 1966. Copyright Bibliothèque centrale du Muséum national d'histoire naturelle (Paris).

FIG. 2. *Arbor sinica Pipa*, in L. Plukenet, *Amaltheum botanicum*, 1703–1704.

named *Crataegus bibas*,[26] called in Chinese *Pi pa xu*.[27]

From these various sources, available in his library, Banks recognised that the Chinese *pipa* and the Japanese *biwa* were a single plant, that this plant had an economic interest and should be imported into England for horticultural purposes, and lastly, that an accurate botanical illustration was still lacking.

What was the response? We know from the note to "An Account of the cultivation of Mespilus japonica (...)"[28] that this tree "was first imported into England from Canton,[29] and placed in the Royal Gardens at Kew, under the auspices of Sir Joseph Banks, in 1787, since which time it has been much propagated and is now to be found in every good collection of exotics in the kingdom". Moreover, in a volume of engravings, published by Banks in 1791[30] from sketches by Kaempfer kept in the British Museum, there is a double plate representing the flower and the fruit of the Japanese medlar (fig. 3) referring to the Latin binomial given by Thunberg. By doing this Banks had given the last touch to 150 years of botanical investigations and paved the way for further researches, theoretical and practical.[31] Furthermore, with the plant at hand, it was possible to study it better: as proof, consider the coloured engraving (fig. 4) from a picture by William Hooker, published in the third volume of the *Transactions of the Horticultural Society of London* in 1820, and the fact that the botanist John Lindley, taking it out of the genus *Mespilus*, would soon create a new genus, *Eriobotrya*, for this plant which — as he still recognised in 1849 — "was deserving of the most extensive culture, both as a plant of ornament and utility.[32] It is still an ornamental, but it can only ripen its fruits after warm summers in Great Britain; however, it is certainly not by chance that it grows so widely today in some parts of Australia. This brief history of the Japanese medlar reminds us of the crucial notion of "acclimatization", which was so important to Banks and which inspired his policy of introduction. As he wrote himself:[33]

"Respectable and useful as every branch of the horticultural art certainly is, no one is more interesting to the public, or more likely to prove advantageous to those who may be so fortunate as to succeed in it, than that of inuring plants, natives of warmer climates, to bear without covering, the ungenial springs, the chilly summers, and the the rigorous winters, by which, especially for some years past, we have been perpetually visited."[34]

Banks may have been dreaming of what could be discovered in China and what could be brought to Great Britain. Even if, in spite of his efforts, the botanical results of Lord Macartney's Embassy (1792–1794) were very disappointing, he would later prepare botanical and horticultural instructions for Dr Clarke Abel, chief medical officer and naturalist of the Amherst Embassy (1816–1817). Meanwhile he had succeeded in having the King appoint William Kerr as botanist and plant collector in Canton.

Why such an interest in China? Maybe through readings like Du Halde's and Grosier's descriptions of the Chinese Empire where one finds interesting chapters on Chinese plants. In fact, in a letter to David Lance dated 18 March 1803, he used the actual words of the first chapter on plants of Grosier's book when he writes that "all fruits grown in Europe are grown by the Chinese". And his idea of exchanging European plants for Chinese ones "because some of their varieties are not so good as ours" may come also from the same source.[35] Besides scientific reasons — climatic, botanical — he might have been influenced by the enthusiastic descriptions of some parts of the Chinese

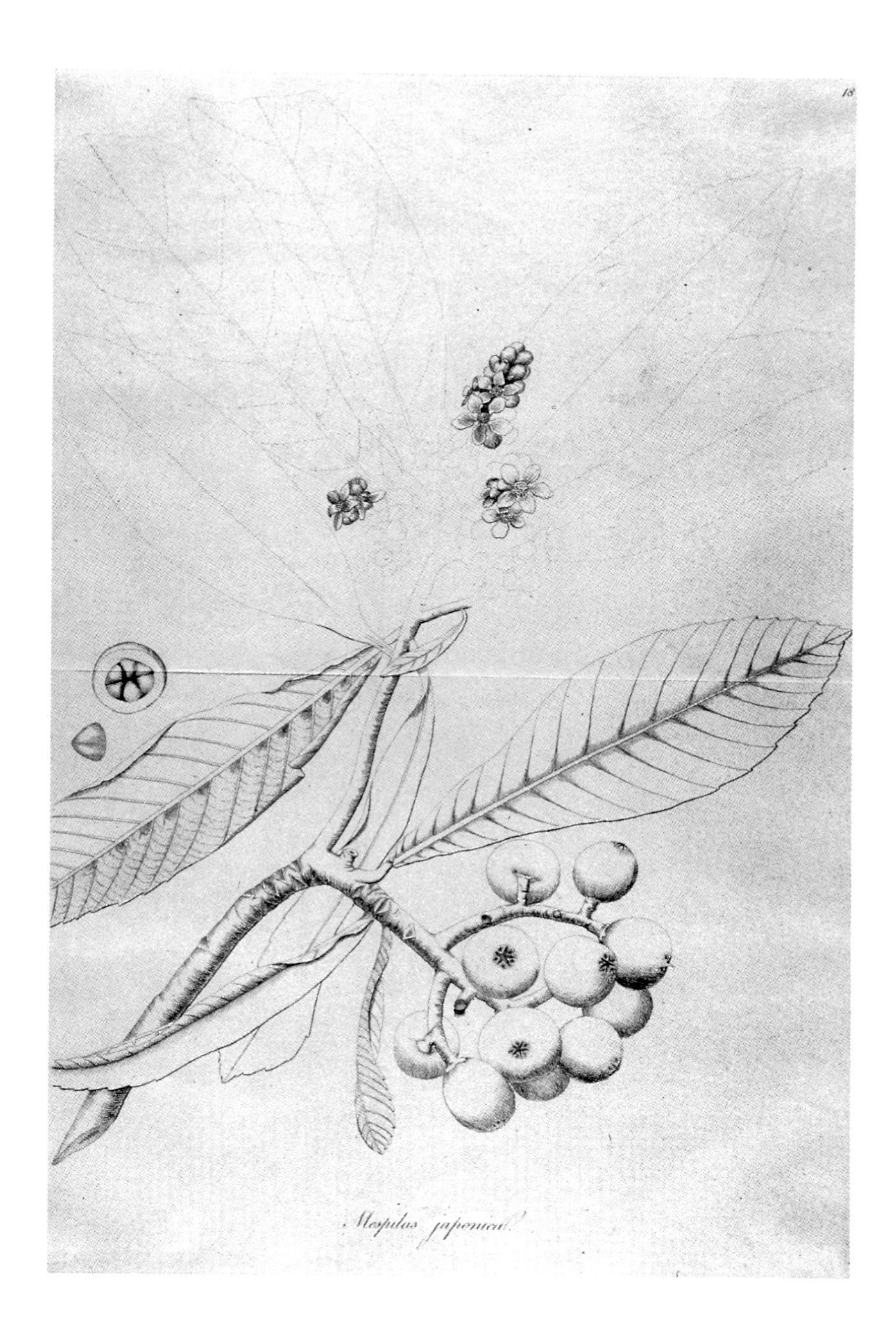

FIG. 3. *Mespilus japonica*, in J. Banks, *Icones selectae plantarum* ..., 1791. Copyright Bibliothèque centrale du Muséum national d'histoire naturelle (Paris).

FIG. 4. *Mespilus japonica* or lo-quat, in W. Bagot, *Transactions of the Horticultural Society of London*, 1820. Copyright Bibliothèque centrale du Muséum national d'histoire naturelle (Paris).

flora by these writers who had used first-hand reports or translations of Chinese sources, done by the Jesuit missionaries. He was also very familiar with more technical texts like the various *Mêmoires* by Le Comte.

We would like to insist on the fact that exploration in China was for Banks not at all of the same kind of the one which was carried on in unknown lands. Besides questioning travellers, sailors and merchants, Banks relied heavily upon his library. It is interesting to note that his "Hints on the subject of Gardening to the gentlemen who attend the Embassy to China" are mainly explanatory commentaries on plants quoted in *Les Nouveaux Mêmoires* by Le Comte. In this way, the persons in charge of collecting plants in the Embassy of Lord Macartney knew what they were supposed to see — and possibly, collect — in different parts of China.(36) He did not limit his researches to literature on China alone. The book on Japan by Kaempfer, with a fifth chapter particularly rich on plants, was always used by him, as was the *Flora Cochinchinensis* of Loureiro, which listed many plants from China with their Chinese names in Portuguese transcription. So he had a good idea of the variety of plants of medicinal, economic and ornamental interest that could be found there, and he prepared lists of plants to collect if possible. On this point, his book of engravings from the sketches by Kaempfer is of particular interest. About fifty plants are represented, most of them identified by reference to the *Flora Japonica* by Thunberg. But a few at the end of the book have only their Japanese name as quoted by Kaempfer. Among the twenty-one plants that, as wrote Banks, "were likely to be more acceptable to his majesties botanic garden than the rest",(37) and that he asked the Embassy to bring back, nine have just their Japanese names without any Latin binomial, eight of which figure in his *Icones* from Kaempfer. In this way the *Icones* appears also as a very practical catalogue for plant collectors. Besides plants, he also asked travellers to look for information about specified cultural techniques. So far as China is concerned, Banks appears to have known well not only what he wanted, but also what he did not know properly and wanted to have investigated because, as he wrote to Lord Macartney in January 1792,(38) "the Chinese appear to me to possess the Ruin of a State & Civilisation in which when in Perfection the human mind has carried all Kinds of Knowledge to a much higher Pitch than the European have hitherto done". One should see here more than exploration; one should think of a real quest. His thoroughness was well rewarded, because the loquat was not the only success: between 1780 and 1817, he is credited(39) with introducing 29 Chinese plants into the Royal Gardens at Kew, among them the now popular *Hydrangea hortensis* Smith (1788), and two peonies, the tree peony(40) (1789) and the double sweet-scented Chinese peony(41) (1805). If we add some 25 more plants selected by William Kerr in Canton and sent to England by various ships of the East India Company, we may note that he was directly involved in the introduction of more than half of the new plants brought back from China during the same period and cultivated at Kew Gardens.

Conclusion

Considering the tremendous change in the tea trade caused by the introduction of the tea plant into India, and considering the fundamental part played by Sir Joseph Banks in this process, one might be tempted to see there the result of an original and specific policy. Actually, we would say that at the beginning it was more likely one specific aspect of a general policy of investigating exotic floras in order to develop new products in England and the dependencies of the British crown. But this transfer of tea from one Asian country, where the access was very difficult, to another which was in the process of colonisation, is certainly the most striking case by reason of its longlasting success and phenomenal political and economic consequences. On this point, I would like to recall the acknowledgement given in 1850 by a Minister of Agriculture of France, where acclimatization of tea plants had been tempted in several places (in Corsica, as we have said, with the help of Banks himself, but also near Paris, Marseille, Montpellier, Foix, Toulouse, Angers) without any success. Baron d'Hervey-Saint-Denis(42) said:

"While we were making the most beautiful projects about, as we said then 'the French tea', our neighbours from England, without noise and theory, were creating huge plantations, not, it is true, in the green and humid meadows of Great Britain, but at the foot of the Himalayas, in a sunny country where the shrub would develop and mature its leaf. The experiment was too rational not to succeed; the success even exceeded all the prospects, to the point that one foresees now the moment when the Anglo-Indian teas will be in competition with those of China on the Asian markets".

And this seems to be a good acknowledgement of a successful Asian policy: the clearsightedness and the perserverance of Sir Joseph Banks, only an advisor at the beginning, had given the impulse which became surely one of the main assets of the success of this policy.

Notes

(1) *Smilax china* L.

(2) A. Franklin, *La vie privée d'autrefois: le café, le thé & le chocolat*, Paris, 1893, 148–9.

(3) H.C. Cameron, *Sir Joseph Banks – The Autocrat of the Philosophers*, London, 1952, 75. Thanks to this success, several naturalists after him (like Charles Darwin), accompanying "naval voyages of Exploration (...) were enabled to collect material and experience which shaped their ideas". (A.G. Morton, *History of Botanical Science*, London, 1981, 324–5.)

(4) E. Mayr, *The Growth of Biological Thought*, Cambridge, Mass., 1982, 867.

(5) Letter Banks to Loureiro, 22 December 1782, Muséum National d'Histoire Naturelle, Paris.

(6) M. le Baron de Cuvier, *Eloge historique de M. Banks, lu à l'Académie Royale des Sciences, le 2 avril 1821*, Paris, 27p.

(7) On the role of botanic gardens in the British imperial policy during the 19th century, cf. L.H. Brockway, *Science and Colonialism: The role of the British Royal Gardens in Empire-Building*, City University of New York, Ph.D., 1977, 299p.

(8) M.-P. Dumoulin-Genest, 'Itinéraires des plantes chinoises envoyées en France: voies maritimes — voies terrestres, St Petersbourg, ville de confluences', 7 éme colloque de Sinologie, Chantilly, septembre 1992, 13.

(9) Broussonet had been introduced to Banks by André Thouin. He went to London in December 1780 to work on the Banksian fish collection (H.B. Carter, *Sir Joseph Banks 1743–1820*, London, 1988, 174). In a letter dated 20 February 1781 to Thouin, Banks writes: "Monsieur le Dr. Broussonet que vous m'avez recommandé me fait beaucoup de plaisir par son agréable et utile conversation; que je vous remercie donc de m'avoir procuré la connaissance d'un ami si utile et si estimable', (Muséum National d'Histoire Naturelle, Paris).

(10) British Museum, London, Add. MS. 80–96. 24–26. Details about letters in British collections quoted in this paper are found in W.R. Dawson, *The Banks Letters*, London, 1958, 965p.

(11) Letter in the British Museum, London, Add. MS. 80–96. 32–33.

(12) Carter, op. cit. (9), 272.

(13) Quoted in ibid, 271–3, which gives a complete account of this document. Other parts of this letter are to be found in Cameron, op. cit. (3), 72–3.

(14) W.H. Ukers, *All about tea*, New York, 1935, vol. 1, 134.

(15) Ibid, 138.

(16) This was one good reason, associated with its high economic value which can explain his great interest for tea. (Cf. Banks's note appended to the letter of 27 December 1788, loc. cit. (13).)

(17) One thinks of the successful introduction of Spanish Merino sheep, cf. Carter, op. cit. (9).

(18) *Eriobotrya japonica* Lindl.

(19) J. Dryander, *Catalogus Bibliothecae Historico Naturalis Josephi Banks*, London, 1798–1800, 5 vols.

(20) M. Boym, *Flora Sinensis*, Vienna, 1656.

(21) E. Bretschneider, 'Early European researches into the Flora of China', *Journal of the North-China Branch of the Royal Asiatic Society* (N.S.) (1880), **15**: 62; L. Plukenet, *Almatheum Botanicum. Opera omnia botanica*, London, 1705, vol. 6.

(22) E. Kaempfer, *Amoenitatum exoticarum politico-physico-medicarum fasciculi v: quibus continentur variae relationes, observationes & descriptiones rerum Persicarum & ulterioris Asiae* ..., Lemgoviae, 1712, chap. 5.

(23) C.P. Thunberg, *Flora japonica – Sistens plantas insularum japonicorum*, Lipsiae, 1784, 419p. Facsimile: New York, 1975.

(24) "Biwa" is a Sino-Japanese term, still in use today.

(25) In a letter to Loureiro, dated 22 December 1782 (Muséum National d'Histoire Naturelle, Paris), announcing that he is sending back his

manuscript for amendment, Banks writes: "If I had entertained any hopes that the Botanists of the age would have received it with candor and have made proper allowances for the difficulties which you have labored under by the want of books during the time you allowed for compiling it I should not have doubted the propriety of publishing it as it is well knowing that the originality & fidelity of the observations contained in it would make ample amends for the faults. But knowing that mankind have not that liberality of sentiment so much to be wished for & observing that 3 parts in 4 of your new genera were constituted of plants well known to botanists both genericaly and specificaly as in the instances of *Priraldus* which is *Dioscorea* & *Columella* which is *Scoparia* I dared not publish it in the shape in which I received it least more blame than praise and more expence than profit should accrue to both author and editor".

(26) J. Loureiro, *Flora Cochinchinensis* (...), Lisbon, 1790, 319.

(27) Which means "pipa tree".

(28) W. Bagot, 'An Account of the cultivation of the *Mespilus japonica* or lo-quat as a fruit-bearing tree, at Blithfield in Staffordshire' + 'Note by the Secretary', *Transactions of the Horticultural Society of London* (1820), **3**: 301.

(29) It came with its local Cantonese name: loquat. One may notice here that if the botanical names — *Mespilus japonica*, then *Eriobotrya japonica* — refer to Japan, the colloquial English term 'loquat' or the French 'bibassier' are of Chinese origin.

(30) J. Banks, *Icones selectae Plantarum quas in Japonia collegit et delineavit Engelbertus Kaempfer: ex archetypis in Museo Britannico asservatis*, London, 1791.

(31) We fully agree with the statement made by H.C. Carter, op. cit (9), 542, "in his day, perhaps no man more than Banks bestrode with such ease and authority the horticultural territory that lay between the dried specimens on the herbarium bench and the crop plants growing in the field or displayed on the market stall".

(32) J. Paxton and J. Lindley, *A Pocket Botanical Dictionary*, London, 1849, 124.

(33) J. Banks, 'Some hints respecting the proper mode of inuring tender plants to our climate', *Transactions of the Horticultural Society of London* (1812), **1**, 21. (Read at the Society on 3 December 1805).

(34) He concludes the report of his own experience on *Zizania* imported from Canada: "Here we have an experiment which proves, that annual plant, scarce able to endure the ungenial summer of England, has become in fourteen generations, as strong and as vigorous as our indigenous plants are, and as perfect in all its parts as in its native climate". He carries on referring to introduced plants: "Though some of these schrubs ripen their seeds in this climate, it never has been, I believe, the custom of gardeners to sow them; some are propagated by suckers and cuttings, and others by imported seeds; consequently the very identical Laurel introduced by Master Cole, and some others of the plants enumerated by Parkinson, are now actually growing in our gardens; no wonder that these original shrubs

have not become hardier, though probably they would have done so, had they passed through several generations by being raised from British seeds". (ibid, 23)

A reaction to that paper, in a letter to Banks, read on 5 December, 1809 at the Horticultural Society, shows the interest among the public of amateurs for these problems of acclimatization "The late Mr. Pollexfen Bastard (...) who had the greatest number of Oranges and Lemons of anyone in this country remarked above thirty years since (what tends to confirm your experiments), that he founded trees raised from seeds and inoculated in his own garden, bore the cold better than Oranges and Lemons imported". (A. Hawkins, 'On some exotics which endure the open air in Devonshire', *Transactions of the Horticultural Society of London* (1812), **1**, 177).

(35) This exchange was made possible through the President of the Company of Merchants in Canton, Puan Khe Qua (Carter, op. cit. (9), 406).

(36) In the same way, when in 1803 William Kerr was sent to Canton, Banks added to his instructions extracts from Grosier's *General Description of China* (Carter, op. cit. (9), 406).

(37) Hints on the subject of gardening suggested to the Gentlemen who attend the Embassy to China. Letter Banks to Sir George Leonard Staunton, 18 August 1792, Linnean Society, Archives.

(38) Carter, op. cit. (9), 291.

(39) E. Bretschneider, *History of European Botanical Discoveries in China*. Leipzig, 1962 [1898], vol. 1, 203–5.

(40) *Paeonia suffruticosa* Andr. = *Paeonia moutan* Sims.

(41) *Paeonia albiflora* Pall. var. *fragrans*.

(42) Baron d'Hervey-Saint-Denis, *Recherches sur l'agriculture et l'horticulture des Chinois et sur les végétaux, les animaux et les procédés agricoles que l'on pourrait introduire avec avantage dans l'Europe Occidentale et le Nord de l'Afrique*, Paris, 1850, 262p.

D. Middleton, 'Banks and African exploration', in *Sir Joseph Banks: a global perspective* (eds. R.E.R. Banks and others), Royal Botanic Gardens, Kew, 1994, 171–176.

13. BANKS AND AFRICAN EXPLORATION

DOROTHY MIDDLETON

Royal Hospital Road, London

We have heard much from this Conference of Joseph Banks's multifarious interests and activities — of Banks the botanist and traveller, of his interest in Australia and of his relations with Europe; of Banks the landowner and of Banks the agriculturist — above all of Banks the President of the Royal Society. It falls now to my happy lot to introduce to you Banks the father of African exploration, from whose teeming brain was born the Association for Promoting the Discovery of the Interior Parts of Africa.

The African Association, as it was then and is now generally known, originated in the discussions of Saturday's Club, one of those gatherings popular with the scientifically-minded gentlemen of eighteenth-century London, meeting periodically to air questions of the day over a good dinner in a comfortable inn. Saturday's Club, of which Banks was agreed to be the "life and soul", consisted of twelve members assembled at the St Albans Tavern off Pall Mall. The African Association was founded at a specially called meeting on 9th January 1788, at which nine members were present, including Banks. Here it was resolved that, "as no species of information is more ardently desired, or more generally useful, than that which improved the science of Geography, and as the vast coastline of Africa, notwithstanding the efforts of the ancients, and the wishes of the moderns, is still in great measure unexplored, the members of this Club do form themselves into an association for promoting the discovery of the inland parts of that quarter of the globe".(1)

Enough has been said of Banks to show how eagerly he sought knowledge and how capable he was of applying that knowledge to practical ends. And he had vision; above all a child of the Enlightenment, he desired to see the boundaries of the known world thrust back, the lot of humanity improved. We are here in one of the great ages of exploration (of what it is no longer the fashion to call "discovery"), an age in which Banks's peculiar talents were most appropriate to the calls made upon them. Cook's voyages had revealed the Pacific, and Banks had shared in Cook's achievement. No wonder he should now turn to Africa, the Continent on which he first set foot in 1771 on the return journey of the *Endeavour*. The few days which were all Cook could allow Banks and Solander to go ashore were enough to convince the two botanists of the richness and beauty of South Africa's flora, and Banks was to continue his interest in the Continent. He was in regular correspondence with James Bruce from the time of the latter's return from Abyssinia in 1774, and he was concerned in the dispatch to the Cape in 1772 of the Kew gardener, Francis Masson, to collect plants for the Gardens. In 1780 Banks was one of a group, including the Duke of Northumberland, which organised the visit of the latter's gardener,

William Brass, as a plant collector to the Gold Coast. African exploration was not, however, merely a personal fancy of Banks's; we have arrived at a point in history when Europe was beginning to be interested in the vast land mass lying so close to its shores, and yet barely known beyond its coastal regions. These regions were first visited by the Portuguese in the fifteenth century, partly in a spirit of adventure, partly in search of trade, partly (as may be shown by the stone crosses planted at landfalls made by the caravels of Prince Henry the Navigator) in a desire to spread Christianity — regions later to be exploited and brutalised by the infamous slave trade. By Banks's day, humanitarian influences had begun to have an effect, and, with the great movement for the abolition of slavery and the slave trade, a vision arose of fairer and more equitable commerce between black and white which could be served by a better knowledge of Africa's "interior parts".

They say a man is known by the company he keeps, and Banks's company was wide, including men with different interests and from different classes. He was of course especially at ease in the congenial atmosphere of Saturday's Club to which belonged politicians, landowners and Fellows of the Royal Society, not to mention a bishop, a doctor and a lawyer. The outlook of these men was liberal: they supported the abolitionist campaign and opposed the war with the American colonies. *Primus inter pares* in such company, Banks could inspire and lead — no doubt he could on occasion dominate — but it may be noted that at no time did the African Association have a President or Chairman. Banks was appointed Treasurer and Henry Beaufoy was the Association's Secretary from its inception until his death in 1795. Beaufoy was a Quaker, a Fellow of the Royal Society and a Member of Parliament with contacts on all useful fronts. In addition he had a gift which Banks, with all his talents, lacked — he had a fluent pen and recorded the affairs of the Association with a charm and facility not often found in the records of Learned Societies.

Membership of the new Association was recruited from the exclusive social class to which Banks himself belonged, and must have been highly exclusive; it never exceeded the figure of 109, reached in 1791. The subscription of five guineas a year was the Association's sole source of income, bringing in about £300 a year — or a total of some £13,000 altogether during the 43 years of the Association's lifetime. Of this some £9000 would have been spent on exploration, the rest going on administrative expenses. There was never a deficit, and the Association wound up its affairs with a credit balance of over £700.(2)

I can hardly find better words in which to sum up the nature of the men who founded the Association, than in quoting Robin Hallett, Editor of *Records of the African Association*:

"They were conscious of the remarkable progress achieved in their time in many different fields — not least in that of geographical knowledge, they looked at the map of Africa and saw how much of it lay blank. Those empty spaces, those marks of ignorance seemed, as Beaufoy was to write, a "stigma" and a "reproach". They were convinced that it would be both practicable and useful to extend the field of knowledge. Africa might be shrouded in mystery, but who could tell what its great interior might contain — new products, new plants to delight the botanists and profit the agriculturist, new opportunities for trade, new markets for the growing industries of Britain, new facts about strange people to satisfy the enquiring mind? Curiosity, then, one may say, was their motive but it was a curiosity controlled by rational minds and directed to practical ends".[3]

The Association founded, it became necessary to decide which "interior parts" were to be explored. James Bruce was about to publish an account of his travels in Abyssinia; André Sparrman, William Paterson and other botanists had already published theirs on South Africa. The Association decided to take the field in the vast area of north and west Africa comprising the Sahara and bounded on the south by the Niger, the river of which mediaeval Arab travellers had left some account. Across the forbidding desert for centuries had plodded the caravans loaded with the "*Golden Trade of the Moors*"[4] Routes were well known to traders, between the Atlantic and the Mediterranean coasts of the continent, but jealously guarded from outsiders. Any exploration of this region would include the Niger, on which was known to stand the legendary city of Timbuctu. Here indeed was a field for exploration, and the Association prepared to dispatch what it quaintly called "geographical missionaries".

Later generations have tended to regard the quest for the Nile sources as the focal point of African exploration, but Banks had sent Mungo Park to the Niger before Burton or Speke were born. Something was known of the river in its middle reaches, but little up or down stream. "The course of the Niger", wrote Beaufoy, "and even its existence as a separate stream are still undetermined".[5] Some, speculating from their armchairs, linked the Niger with the Congo, some with the Nile; some maintained that it rose in Lake Chad. And which way did it flow? From east to west as alleged by mediaeval Arab geographers or (more probably) from west to east? The Association determined to find out and at the same time to unveil the mysteries of Timbuctu to which no European traveller had penetrated — as it were, to "put a man on the Moon". No salary was paid to the first three explorers sent out by the Association — it would seem that the experience and the glory with which success would crown them would be sufficient recompense for their trouble! From Park onwards however a salary was paid.

The first "geographical missionary" to be appointed was Simon Lucas, who, as a boy, had been captured by Barbary pirates and forced to spend three years as a slave in Morocco. On his release Lucas had the good fortune to be sent back to Morocco as Vice-Consul and Chargé d'Affaires. Sixteen years later he obtained a post as Oriental Interpreter at the Court of St James's and was instructed to join the quest, proceeding south-west out of Tripoli with a view to crossing the Sahara and striking the Niger. The next appointment owed much to Banks's knack for making contacts with a wider circle of acquaintance than that which frequented the premises of the Royal Society or dined at the St Albans Tavern. This was the American John Ledyard who had sailed with Cook on the third Pacific voyage and on his return had set off to walk across Russia into Siberia and thence to find his way to Nootka Sound on the north-west coast of America. He was forcibly deported to England where he sought out Banks, who gave him an introduction to Beaufoy. The latter was "at once struck with the manliness of his person, the breadth of his chest, the openness of his countenance and the inquietude of his eye"[6], and forthwith offered him the chance of exploring Africa. When asked how soon he could set out, Ledyard replied "tomorrow morning", and in little more than a month arrived in Alexandria on his way to Cairo, where he was to seek out a west-bound caravan across the Sahara on a more southerly route than that recommended to Lucas. Alas, Ledyard died in Cairo from a bilious compliant, allegedly the consequence of vexatious delays, and Lucas was thwarted by rumours of inter-tribal warfare from leaving Tripoli.

The next "geographical missionary" to be employed was an Irishman, Major Daniel Houghton, directed to seek the Niger from the opposite direction. He set off from the Gambia river early in 1790, but disappeared on the way — he was probably murdered but no details became known. It was not until four years later that success was at last achieved, by Mungo Park — the "Jewel in the Association's Crown", and one of Africa's most famous explorers. Park was introduced to the Association by his brother-in-law, John Dickson, a nurseryman sufficiently well up in botany to be on friendly terms with Banks. Park, a Scottish doctor with a passion for travel, was the first of the Association's men to receive a salary — 7s 6d a day in England, 15s a day in Africa. He set off from Karantaba on the Gambia on 2nd December 1795, carrying "a few changes of linen, an umbrella, a pocket sextant, a magnetic compass and a thermometer, two fowling pieces and two pairs of pistols".[7] He wore a large hat in which he kept his papers. He was accompanied by a servant who also acted as an interpreter in Mandingo, a "sprightly youth" called Demba and four other local men, a horse and a donkey. After terrible hardships and robbed of nearly all he had, he reached the Niger at Segu in modern Mali and on 20th July 1796, he saw "with infinite pleasure the great object of my mission, the long sought for and majestic Niger, glittering in the morning sun, as broad as the Thames at Westminster, and floating slowly to the *eastward*".[8] In 1805 the British Government, taking up the running from the Association, sent Park back to the Niger in command of an elaborate expedition, attended by a rabble of ill-disciplined soldiers. The death toll was almost total (mostly from fever), and Park and the last of his European companions were drowned in the Bussa Rapids south of Timbuctu in a desperate attempt to trace the river's course. It was not until 1830, 10 years after Banks's death, that the Lander brothers sailed down the river to where it flowed into the great Delta on the Bight of Benin.

The Association nearly scored a double on the Niger. In the spring of 1796, while Park was on the Gambia preparing for his journey, a determined and well qualified young German, Friedrich Hornemann, was leaving Tripoli with the Association's orders in his pocket to seek the Niger from the opposite direction. Hornemann had orginally approached the Association with a strong recommendation from J.F. Blumenbach, Professor of Medicine at Göttingen University and one of the Banks's most respected friends. Britain and France being at war, he might have had difficulty in crossing France on his way to Egypt, had not Banks (as usual) had a useful friend in the right place, thanks to those close contacts with Europe of which Professor Crosland has spoken. In the previous year Banks had done a good turn for Jean Charretié, French Commissioner in London for the Return of Prisoners of War, causing to be restored to the French scientist, J.J. de Labillardière, the important botanical collection confiscated at the time of the latter's capture.[9] Operating from Tripoli, Hornemann made a careful reconnaissance as far as Marzuq in the Fezzan, returning to Tripoli to report before setting out in earnest for the Niger from Marzuq on 6th April 1800. He was never seen again. It was not until 16 years later that reports were received of his death, probably of dysentery, within a day's journey of his goal.

It was some years later that the Association dispatched Johann Ludwig Burckhardt, who was to be their last man directed to the Niger and also the last to be introduced by Banks. It would seem that much had been learnt since the Association's early days, when the breadth of Ledyard's chest, the "inquietude" of his eye and his willingness to start at once for Africa had so impressed Beaufoy. The Association's view now extended beyond the geography of a river to comprise an appreciation of the Arab world as a whole, and Burckhardt's brief was far wider than that of his predecessor. Burckhardt was a Swiss who had studied at Göttingen and Leipzig, and who came to England in the hope of securing an appointment in the English Foreign Service. Disappointed in this ambition, he visited Banks, to whom he, like Hornemann, had an introduction from Professor Blumenbach. Unlike Ledyard who had been admired for his eagerness to be off "tomorrow morning", a careful programme was made out for Burckhardt, who was to spend several years in the Near East on a salary of 10s 6d a day, learning Arabic and becoming familiar with the language and way of life of the regions he visited. The only link between them was that both were put forward by Banks — Ledyard associated with the glorious days sailing the Pacific with Cook, Burckhardt with the more sophisticated haunts of later times. Burckhardt arrived in 1809 in Syria where he spent three years learning Arabic and travelling around. Arriving in Cairo, he found, as had Ledyard, that there was no caravan leaving to cross the desert in the Niger direction; he therefore set about devising an alternative programme. He travelled south in Nubia, carefully recording all he saw and learnt, then, turning east, he crossed the Red Sea and made his way to Mecca disguised as an Arab merchant. Back in Cairo two years later, he was to share the fate of the impulsive Ledyard, dying of some internal complaint in 1817. His records, carefully kept, were assembled and published by the Association between 1822 and 1831. His collection of Arabic manuscripts was bequeathed to Cambridge University Library.

By the time of Banks's death in 1820, the resources of the Association were proving inadequate to any further penetration of Africa. Undeterred by the failure of Park's second expedition, the British Government had now pursued its programme in Africa and secured promising results from the journeys of Clapperton and the Landers during the early part of the nineteenth century. The Association's swan-song was to sponsor the first attempts in modern times to trace the course of the Nile, despatching up stream in search of the river's source three gallant but abortive attempts — led by Captain J Gordon, RN, in 1822, by the Frenchman Linant-de-Bellefonds in 1827, and by Henry Welford in 1830. In 1831 the Association's affairs were wound up and it merged with the newly founded Geographical Society of London (later the Royal Geographical Society). The exploration of the globe which has ever since been the business of the RGS therefore owes much to Banks. Not only did the African Association prove the usefulness of independent amateur bodies in the field of discovery, but it put in the forefront of its aims (where the RGS has kept it) the paramount importance of extending geographical knowledge. As the American geographer John Kirtland Wright has put it: "The history of geographical discovery is a fabric into which may colourful threads are interwoven, threads of adventure and hardship, of military conquest and political intrigue, of commercial enterprise, of theoretical speculation, and of hard scientific thought".[10] We could all extend this list and I for one would add the desire for knowledge, for attainment of what lies beyond the horizon — to stand in spirit with Mungo Park when at last he saw the "long sought for majestic Niger, glittering in the morning sun... and flowing slowly to the eastward".

It is not in a mere figure of speech that I introduced Sir Joseph Banks, who sent him there, as "the father of African exploration".

Notes

(1) R. Hallett (ed.), *Records of the African Association 1788–1831*, Edinburgh, 1964, 46.

(2) Ibid., 35.

(3) Ibid., 15.

(4) E.W. Bovill, *The golden trade of the Moors*, 2 edn., Oxford, 1968.

(5) Hallett, op. cit. (1), 44–5.

(6) Ibid., 55.

(7) Quoted in C. Lloyd, *The search for the Niger*, London, 1973, 34–5.

(8) Quoted in K. Lupton, *Mungo Park: the African traveler*, Oxford, 1979, 76.

(9) E.W. Bovill, *Missions to the Niger*, Cambridge, 1964, vol. 1, 15.

(10) J.K. Wright, *Human nature in geography*, Cambridge, Mass., 1966, 24.

G. Williams, ' "The common center of we discoverers": Sir Joseph Banks, exploration and Empire in the late 18th century', in *Sir Joseph Banks: a global perspective* (eds. R.E.R. Banks and others), Royal Botanic Gardens, Kew, 1994, 177–191.

14. "THE COMMON CENTER OF WE DISCOVERERS" SIR JOSEPH BANKS, EXPLORATION AND EMPIRE IN THE LATE 18TH CENTURY

GLYNDWR WILLIAMS

Queen Mary and Westfield College, University of London

Whatever the aura of prestige which surrounded Sir Joseph Banks in the multifarious activities of his middle and later careers, for some he always remained the "Mr Banks" who had sailed with James Cook on the *Endeavour* voyage. It is hard to overestimate the importance of those three years in the development of Banks's character and outlook. We have some inkling of it in his journal of the voyage; more in his restrospective chagrin after his refusal to sail on the second voyage on any terms but his own; and more still in the care he took to repair relations with Cook after the return of the *Resolution* in 1775. By now there had taken place what J.C. Beaglehole has called "a curious development": Banks had become "a sort of superintending elder brother in relation to some of the concerns of Cook".(1) During the explorer's long absences Banks became, as it were, his spokesman — in a way the custodian of his reputation. The events of the third voyage showed this quite clearly. It was Banks to whom Gore entrusted the care of his child if he did not return; Banks to whom Clerke wrote his last, sad letter; Banks who saw all the official dispatches from the voyage; and, most decisively, Banks who became "the principal adviser and co-ordinating authority" for the publication of the official narrative of the voyage.(2) It is little wonder that in late 1780 James King, who had brought the ships home after the deaths of Cook and Clerke, wrote to Banks that "I look up to you as the common center of we discoverers".(3)

During the 1780s Banks — by now Baronet, President of the Royal Society, adviser of cabinet ministers, patron of the sciences on an international scale —did much to help the careers of Cook's men. But he did more: he became the promoter of enterprises associated with Cook's discoveries and (as far as the first voyage was concerned) with his own. Breadfruit from Tahiti, settlement at Botany Bay, the fur trade of the northwest coast of America, the southern whale fishery, all attracted Banks's attention. Away from Cook's old stamping grounds, we find Banks in his garb of botanist preaching the gospel of plant interchange to help develop imperial self-sufficiency. He was involved in attempts, sometimes haphazard, occasionally bizarre, to transfer cotton seeds from India to the Caribbean, tea plants from Canton to Bengal, cochineal insects from South America to Madras — to list only a few of his activities. Banks was a one-man correspondence centre, academic institution, finance house. In the period of national reorganisation after the War of American Independence he was a central figure in renewed imperial expansion.(4) This paper will take as case-studies just two of the many regions which attracted Banks's attention in these years, the North Pacific and West Africa.

This was the period when Banks bore important testimony as to the suitability of New South Wales for settlement. There was even a hint that he might return there in person, though the letter to him in July 1783 by James Matra, his former shipmate on the *Endeavour*, is more important for its revelation of schemes and projects in the air.

"I have heared a rumour of two plans for a settlement in the South-Seas; one of them for [New] South-Wales, to be immediately under your direction, and in which Lord's Sandwich, Mulgrave, Mr Colman, and several others are to be concerned. The other a distinct plan, in which Sir George Young, and Mr Jackson, formerly of the Admiralty are the Principals. I have met these stories in several romantick shapes...".(5)

Within the month, Matra had sent his own plan for a settlement in New South Wales to Lord North, noting Banks's "high approbation" of it.(6) The complicated sequence of events which intervened before such a settlement was actually established has been dealt with before, not least in Alan Frost's presentation to this conference of Banks's vision of New South Wales as an "Antipodean Europe".

Banks's thinking about the Pacific at this time was clearly influenced by his first-hand experience on the *Endeavour* voyage, but also by the discoveries of Cook's third voyage, of which he had early sight as he helped to prepare the journals for publication. A note in the memoirs of Sir John Dalrymple describes "an Intended Expedition into the South Seas [in 1779] by private Persons in the late war", and continues:

"I have sometimes talked of the above project to Sir Joseph Banks, who observed, that since the discovery of the Sandwich Islands by Captain Cook, in his last voyage, such adventures are become much more easy; because, in these islands, the adventurers will find places of refuge for their ships, provisions for their crews, strong stations in which to lodge their plunder, from whence they may return to get more, and inhabitants in the islands to assist their seamen in sailing either to the east or to the west".(7)

Those sailing east would come, as Cook had done, to the northwest coast of America, where his crews had traded sea otter skins which fetched high prices at Canton on the homeward voyage. News of this in the unofficial narratives, and then in the official account, of the voyage led to a rush of trading expeditions to that remote coast. Already, in 1783, Joseph Billings, who had been with Cook on the third voyage, laid before Banks a scheme for a further expedition,(8) but it was 1785 before the first trading vessel, commanded by James Hanna, reached the coast, and that was fitted out in India. The next year six ships were on the coast: four from India or Canton, and two which had sailed the fourteen thousand miles from Britain. Banks had early news of Hanna's venture. Charles Blagden had seen a copy of Hanna's journal by September 1786, and a note in Banks's hand has some of the details of the voyage and a rough (if exaggerated) indication of its profitability at Macao — "brought 500 Skins Hong Merchants to bid 70,000 dollars had made a Contract for a Supply in Future".(9)

In different ways Banks was involved with two of the expeditions of 1786. James Strange was in overall command of one of these. It consisted of two vessels from India, the *Captain Cook* and the *Experiment*, owned by David Scott, a free merchant of Bombay, but operating in close association with the East India Company. The Company, together with the moribund South Sea Company, held sole rights of British trade east of the Cape of Good Hope. Among Banks's papers are the detailed sailing instructions given to Strange, which were sent to Banks for his approval, and also "Extracts from the Letters of Mr and Mrs Strange", endorsed with Sir John Dalrymple's name.[10] Another letter, probably to David Scott, from William Hunter on the *Experiment*, was copied by Banks. It details the adventures and misadventures of the voyage to the northwest coast, from which the ships returned to Macao on 14 November 1786:

"to the happiness of all on board most of our men dying of the scurvy we have been laying here a week and are not yet determined what we are to do there being a great many furrs in the China market and the China men say they have plenty there are now 6 ships on the Coast all for furrs you may suppose the Quantity they will get when we got between 4 and 500 skins and only one ship had been there before us and she was not at Nootka for what reason I cannot say we got more skins there than at any other place on the coast our ship goes back if it is only to fetch the doctor I shall leave here and go back to Bombay for 45 Rupees a month will never make a Fortune".

In England, Banks was one of the first to hear that the potential of the northwest coast fur trade was perhaps more limited than had at first been thought. In terms of exploration, however, intriguing possibilities has been opened up by Strange's discovery and naming of Queen Charlotte Sound in latitude 51°N., shown in a rough sketch map by Banks leading inland north of (present-day) Vancouver Island.[12] This opening had not been seen by Cook in 1778, and realisation was dawning that the great navigator's dismissal of the inland straits reported to have been discovered by Juan de Fuca in 1592 and Bartholomew de Fonte in 1640 might have been premature.[13]

The two ships which arrived on the coast in 1786 from England were commanded by Nathaniel Portlock and George Dixon, both of whom had been there with Cook on his last voyage. The expedition had been fitted out by an association of merchants formed the previous year in England, and named the King George's Sound Company. Headed by Richard Cadman Etches, it obtained — not without difficulty — a licence from the East India Company to trade on the northwest coast.[14] Evidence from the Etches family and from Portlock shows the close connection between the venture and Banks. Richard's brother, John, recalled that Banks was a leading patron of this scheme "for prosecuting and converting to national utility the discoveries of the late Captain Cook, and for the establishing a regular and reciprocal system of commerce between Great Britain, the North-west coast of America, the Japanese, Kureil, and Jesso islands, and the coast of Asia, Corea, and China".[15] In his published account of the voyage, Captain Portlock recalled how Banks was one of those who came on board the ships as they were fitting out, and named one of them the *Queen Charlotte*.[16] We know that Banks loaned Portlock one of Cook's logs, delivered by Gore,[17] and correspondence between Richard Cadman Etches and Banks shows the importance of Banks's support in other ways. A letter from Etches in

March 1785 indicates that the possibility of using the expedition to open trade with Japan was Banks's idea, and this is strengthened by a letter from Banks to his Swedish colleague, the naturalist Carl Peter Thunberg, in which he asked as a matter of urgency for further information on this subject.[18] The will-o'-the-wisp of the Japan trade attracted Banks, as did the potential of the China market. As he explained to Henry Dundas, all means possible ought to be taken to encourage British imports there and "diminish at least if not annihilate the immence debit of silver which we are annualy obligd to Furnish from Europe".[19]

When planning later ventures to the northwest coast Etches referred to Banks's "repeated Acts of Friendship", and went on to acknowledge

"Your patronage I have experienced Your assistance I have found of the utmost consequence, the past enduces me to solicit your future endeavors to complete our Views — Protection from Government is what we presume to hope for —You Sir I am sure can adopt an Idea in what manner this may be done with propriety".[20]

In correspondence and meetings with Banks in 1788 Etches sketched his plans for the future. He had outfitted another venture to the northwest coast, under the command of James Colnett and Charles Duncan, which had arrived there the previous summer. On board, at Banks's recommendation, was Archibald Menzies as surgeon and botanist. Etches' objective was more ambitious than a series of separate trading voyages, for despite the failure of the Portlock/Dixon expedition to set up a permanent base on the coast ("their Powers of Government were not competent to the task", he told Banks), Colnett and Duncan had been instructed to establish at least one trading post. This might be located on the "new island" (the Queen Charlotte Islands) discovered by Dixon in 1787. "Our intention", Etches insisted, "is to adopt a permanent system of Comerce direct from this Country to the N.W. coast and from thence to the Asiatic Coast and Islands.[21] The post or posts might be manned, Etches suggested, by convicts and guards from Botany Bay. Only a few months earlier, James Strange had put to the Governor of Bombay "the practicability of forming a Settlement on the North West coast of America", manned by an establishment of the "unfortunate Criminals" from New South Wales.[22] Banks's doubts about the cost of such an arrangement did not deter Etches, who went on to suggest that a small naval vessel might also be sent to the coast to protect the trade and carry out survey work. This would open up the fur trade with the natives of the interior, and on "the opposite shore" with "the Japan Islands".[23] In a further letter Etches tried to retain Banks's interest by passing on details he had just heard from crew members of William Barkley's trading vessel, the *Loudoun* (or *Imperial Eagle*) about their discoveries in 1787.

"They left the Sound [Nootka] and Coasted about 40 leagues to the Southward, frequently Anchoring among the Islands (to Traffic) till they came to a large strait...From the whole Conversation I had with them their appears evident probability that this is the strait laid down by Juan de Fuca and from the depth of water, bredth, strong tide and the trending to the N.E. we may presume that the Western Sea laid down by de Fuca, de Fonte, and Aguilar hath an existance, and the whole now laid down as Continent is nothing but Islands...From this and evry other information we have been able to obtain, I am fully perswaded that their is evry probability of the whole Country from the Strait to the latitude of 60N being an immense Chain of Islands and from what we hitherto know of its Traffic must prove of imense Value."[24]

Within weeks of Etches passing on news of this discovery to Banks, one of his own vessels, the tiny *Princess Royal*, commanded by Charles Duncan, entered the strait and mapped its entrance. News of this second apparent confirmation of the existence of Fuca's elusive strait did not reach England until 1789; but in the meantime Dixon arrived home in September 1788, and sent Banks a memorandum in which he related his explorations of the previous year to the supposed discoveries of Admiral de Fonte in 1640. He described how he had "Surveyed all a long the Coast betwixt Woody Point and Cape Edgecombe, and thinks that it has very much the Appearance of the Archipelago Islands, said to be seen by Admiral de Font. There are many deep Inlets; the Main he thinks he has not seen". Dixon also drew attention to the international aspect of all this. "The Russians are settled in Cooks River and at Beerings Bay. The Spaniards are making Settlements somewhere a little South of King Georges Sound [Nootka]. The French Ships on discovery [La Pérouse] have been at Macao, sold some skins, and are gone to the Northd."[25]

By the late eighties the northwest coast was attracting the attention of geographers, merchants and, finally, governments. There, despite the insistence to the contrary of Cook in 1778, it seemed that a strait leading deep into the interior might yet be found. Its furs might provide the missing link needed to make viable a northern network of trade encompassing China and Japan. Foremost among the advocates of renewed exploration along the northwest coast was Alexander Dalrymple.[26] The relationship between Dalrymple, at this time hydrographer to the East India Company, and Banks was a long-standing one. It was to Banks that Dalrymple in 1768, swallowing his disappointment at not being given command of the Pacific voyage, had presented a pre-publication copy of his *Discoveries made in the South Pacific Ocean. Previous to 1764*. Through the Royal Society and their mutual interests Banks and Dalrymple knew each other well, though despite the hydrographer's help with the charts in the publication of the official account of Cook's last voyage, and evidence that he had the run of Banks's library, it is doubtful if the two men were particularly close.[27] Banks had been sceptical of Dalrymple's theories about a southern continent, and more recently they had been at odds over the choice of Botany Bay. Nor can Dalrymple's criticism of Cook's opinions on the geography of the northwest coast — "I cannot admit of a Pope in Geography or Navigation" he wrote at one point[28] — have made pleasant reading for Banks. There is, too, an obscure episode at the end of 1787 in which Dalrymple responded to some proposal or other from Banks with "a Plump refusal".[29] Nevertheless, Dalrymple was the key figure in these years as he collected and interpreted reports from every possible source on the geography of the northwest coast, and then urged action on government ministers and on the boards of the East India and Hudson's Bay companies. Here again, Banks may have provided a link, for the hitherto closed archives of the Hudson's Bay Company were opened to Dalrymple by the Company's Governor, Samuel Wegg, a close associate of both Banks and Dalrymple at the Royal Society.[30]

Central to Dalrymple's thinking was his conviction that the distance between the Pacific coast of North America and the interior lakes of the fur-traders was less than the current estimates. In terms of knowledge, if not action, Banks had anticipated Dalrymple in this area. As early as 1781 Banks and Solander had listened to the Canadian fur-trader, Alexander Henry (the Elder), explaining that he thought the knives found by Cook among the Indians of the northwest coast were the ones traded by him in the Athabasca region in 1776.[31] On his return to Canada, Henry sent Banks the plan of "A Proper Rout, by Land, to Cross the Great Contenant of America from Quebec to the Westermost Extremity". From the Athabasca country, only recently reached by the fur traders, a "Great River" flowed down to the Pacific — not more than thirty days' travel to the coast. There a trading post could be established, and a small vessel built to trade and explore.[32] Banks's response to this is unclear, but this brief and (as far as we know) only meeting with Banks impressed Henry sufficiently for him to dedicate his *Travels and Adventures in Canada and the Indian Territories* to Banks — more than a quarter of a century later. In 1784 a similar scheme was put to Banks by George Dixon, whose proposal was turned down by Banks on the grounds that the recent war had left the country in an unsettled and dangerous state, and that in any case he had heard that the Americans were preparing a similar undertaking.[33]

Away in the northwest it was indeed an American, Peter Pond, though working for the British North West Company, who was the most enterprising of the interior explorers; and Banks was an important source of information about his activities. A scribbled slip in Banks's hand notes that in 1788 a visitor to Revesby brought news about an attempt by Pond to cross the continent, and that "he had before traveld till he met the Tide in a river which ran to the west & supposes he then was within three days Journey of the Sea".[34] More detailed information reached Banks in January 1790 in the shape of a letter from J. Marvin Nooth of Quebec, one of his botanical correspondents. Nooth's letter of November 1789 related how

"a very singular Person of the Name of Pond is arriv'd at Quebec. This man has been some years in the western parts of America on a trading Expedition with the Indians, & positively asserts that he has discover'd an immense Lake [Great Slave Lake] nearly equal to Great Britain that communicates in all probability with Cooks River or Sandwich Sound. In the River which was form'd by the Water that was discharg'd from this Lake he met with Indians that had undoubtedly seen Cooks Ships and who had with them a variety of European Articles evidently of English manufacture".[35]

Nooth added that Pond had presented a map of his discoveries to Lord Dorchester, Governor of Quebec. More information about the map was given in another letter from Isaac Ogden, a Crown official in Quebec, which his father in London sent Under-Secretary of State Evan Nepean, an associate of Banks and Dalrymple. The original letter is in the Colonial Office files, and there is a copy among Banks's papers.[36] The new information from Quebec seemed to support the argument of Dalrymple, illustrated in his "Map of the Lands of the North Pole" of 1789 (of which Banks had a copy) that a passage through or round North America to the Pacific might yet be found.

In London speculation and plans about the northwest were brought to an abrupt halt when at the beginning of February 1790 news arrived from Madrid of the Spanish seizures of British vessels and property at Nootka Sound the previous summer. Among the initial responses of the British government was a plan to send a naval expedition to the northwest coast, commanded by Cook's old shipmate, Henry Roberts. The ships were to go by way of Port Jackson, and there take on board thirty men selected from the New South Wales Corps, convicts and overseers, to establish a settlement "for the assistance of his Majesty's subjects in the prosecution of the fur trade from the N.W. coast of America".[37] It was a version of the plan suggested by Etches to Banks in July 1788, and supplemented by a letter of October 1789 from George Dixon who again urged on Banks the advantages of the Queen Charlotte Islands as a base.[38] It was this general region, indeed, rather than the sensitive area of Nootka, which the government seemed to have in mind; for one of the ships under Roberts's command was directed to land its settlers, if possible, at Queen Charlotte Sound in lat. 51°N.[39] Banks's advice was sought in the matter of what trade goods the settlement would need, and here Banks turned to Menzies, just back from the northwest coast in the *Prince of Wales*. Menzies's long list of recommended goods would cost, Banks estimated, £6,800.[40] This was a not inconsiderable sum, but in language unusual for a government minister, Secretary of State Grenville insisted that it was "very desirable not to stint that part of the service", and warned against any "ill-judged economy".[41] It was reasons of state, not cost, which led to the abandonment of the Roberts expedition as the crisis with Spain intensified, and a general mobilisation of the fleet was ordered.

Following the Nootka Sound Convention of October 1790 the naval expedition was reinstated, but with different instructions and a different commander. George Vancouver carried no orders about new trading establishments; his task was partly exploratory, partly diplomatic. Banks's association with the expedition went well beyond the appointment of the botanist, Archibald Menzies. A tantalising reference in a reply to Banks in December 1790 by Lord Auckland at The Hague, seems to indicate some wider interest by Banks in the Nootka area — "These are imperial words & worthy of our Sovereign", wrote the diplomat.[42] In the same month a letter from Nathaniel Portlock shows that Banks had secured his appointment to command of Vancouver's consort vessel on the expedition, the *Chatham*. For health reasons Portlock declined the command, though as he told Banks "The fear of offending you Good Sir in this my determination gives me much uneasiness".[43] At Grenville's request, Banks sent detailed instructions as to how Vancouver's survey should be conducted. Drawn up with the help of James Rennell these formed "the most detailed guide for scientific and marine exploration ever set out in the eighteenth century".[44] Over and above the technical instructions on surveying methods there was the over-riding urgency of determining whether any of the openings recently discovered along the northwest coast led into the interior. This, the sceptical Vancouver was told, was "one of the principal objects of your Survey".[45] Banks's supplementary instructions to Menzies were mostly about his botanical duties, but there was an additional dimension. By now it was known that Vancouver's expedition, unlike the original one under Roberts, would not be concerned with establishing trading settlements. Even so, Banks instructed Menzies that he was to pay particular attention to the climate, soil and natural products, in case it should at "any time hereafter be deemed expedient to send out Settlers from England".[46] Something rather more substantial than trading posts seems, however distantly, to be envisaged here.

The role of Banks in the surge of British enterprise in the North Pacific after Cook's final voyage varied from knowledgeable encouragement to direct intervention with government, from the collecting and distributing of information to the supervision of the botanical side of the voyages. It is a role which is sometimes shadowy, difficult to assess in precise terms; for much of what Banks accomplished went unrecorded either in state or private papers. Harold Carter reminds us that many of those with whom he was in contact lived in London, ministers, merchants and navigators "whom Banks could meet in person conveniently and often enough in various ways for business to be done by word of mouth alone".(47) There is a rather different sense in which Banks's role is hard to define. Away from his circle of intimates he worked with men whose objectives and priorities were often different from his own. There were times, undoubtedly, when he was used by them.(48) Only with his beloved botanists, scouring the globe for specimens, could Banks be sure. And however large natural history loomed in Banks's world, for the hard-headed agents of imperial expansion it was not a matter of great moment. Even relations with his intimates among government ministers could bring problems. Almost thirty years after the event, Banks still recalled, and resented, Pitt's insistence that he should use "every possible degree of Economy" in organising the *Bounty* voyage. "This niggardly arrangement", Banks wrote, led directly to the mutiny.(49)

It was perhaps the continuing frustration of influencing but not controlling which led Banks towards the founding of the African Association in 1788. There is something breathtaking, almost outrageous, about the Association's establishment. Nine well-to-do Englishmen, meeting for dinner in a Pall Mall tavern, mulling over one of the most formidable tasks of exploration of the era, decided to form an Association, five guineas annual subscription, "for promoting the discovery of the interior parts of Africa". Among the five committee members, Banks was to be Treasurer and Henry Beaufoy Secretary. The formal "Plan" of the Association was drawn up by Beaufoy with admirable clarity:

"Notwithstanding the progress of discovery on the coasts and borders of that vast continent, the map of its interior is still but a vast extended blank, on which the geographer...has traced, with a hesitating hand, a few names of unexplored rivers and of uncertain nations...sensible of this stigma, and desirous of rescuing the age from a charge of ignorance, which, in other respects, belongs so little to its character, a few individuals, strongly impressed with a conviction of the practicality and utility of thus enlarging the fund of human knowledge, have formed the plan of an Association for promoting the discovery of the interior parts of Africa".(50)

For Banks the Association became a personal vehicle, not for his own aggrandisement, but to satisfy that compelling urge to promote his particular blend of science, trade and empire. He would choose his own men, decide what they were to do, and where they were to go. It was the logical development of Banks's progression since the *Endeavour* voyage. Banks remained Treasurer until 1804, and during his long spell of office missed only one meeting of the Committee, often held at his house in Soho Square. In the absence of any President, Banks was, as Beaufoy's successor Bryan Edward wrote, "the life and soul of the Association".(51) In her paper Dorothy Middleton has analysed the successes and failures of the Association's explorers. Here I shall confine myself to the motives of Banks and his colleagues in the early years of the Association's history.

From the beginning the Association was more than an agency for scientific investigation. The Committee reserved the right to keep its explorations secret, and some indication of the motives which played a part in its foundation are reflected in Banks's papers of 1787. At one moment he was involved in correspondence with Manchester cotton manufacturers about the possibility of establishing cotton cultivation in west Africa; at another he was receiving letters from the indefatigable James Matra, just appointed British consul in Morocco, about caravan routes into the interior.(52) Much of the Associations's early effort was centred on the mysterious city of Timbuktu. In 1789 the Committee pointed out that if a route could be opened by way of the Gambia to Timbuktu, "one of the principal Marts for the supplying of the Interior of Africa...such an intercourse may eventually prove of the greatest importance to the Commercial Interests of Britain".(53) In the first of the Association's published *Proceedings* in 1790 Beaufoy elaborated on the union of national interests and "the still more extended interests of philosophy" which such a route would bring. "The gain which the merchants of England would derive from a similar traffic, conducted as is here proposed, would be such as few commercial adventurer have ever been found to yield."(54) There is an appeal here for political and mercantile backing, and Beaufoy had the knack of clothing his remarks in language acceptable to those large sections of society which had turned against the slave trade. So, in 1792 he wrote that "the people of the inland regions of Africa may soon be united with Europe in that great bond of commercial fellowship which the mutual wants and different productions of the other countries of the globe have happily established".(55) The more down-to-earth aspects of the Committee's interests were well represented in Sir John Sinclair's recollection of the occasion when

"the African Club dined at the St Albans Tavern. There were a number of articles produced from the interior parts of Africa, which may turn out very important in a commercial view; as gums, pepper, etc. We have heard of a city where Major Houghton, our geographical missionary, is going, called *Tombuctoo*: gold is so plentiful as to adorn even the slaves; amber is the most valuable article. If we could get our manufactures into that country we should soon have gold enough....".(56)

To Timbuktu was added the equally beguiling prospect of another unknown city, "Housa". It is "one of the first cities in the world", Matra told Banks as he described "Manufactures of different kinds, especially *Linens of exquisite fineness resembling Cambric*, and in such abundance that in one day you may procure as much as would load *fifty large ships*".(57) Even the hoary old legend of a great inland sea was resurrected, this time by the usually sensible Beaufoy.(58) To fit out individual explorers was one thing, and well within the financial capabilities of the Association; to secure the trading advantages Beaufoy wrote about so eloquently was something else, and needed government support. In May 1793 Beaufoy told Banks that "several merchants of great wealth in the City had offered to vest a large sum in the adventure to Bambouk, as soon as they were assured that the countenance and protection of Government would be given to the Plan".[59] In June the Committee wrote to Dundas asking for the appointment of a British consul to Senegambia who might stamp the seal of official protection on the route to the interior by way of the Gambia, and also ease access to the gold mines of Bambuk. Such a move would bring Britain "the trade which is now carried on by the Barbary States to the Inland Nations of the Continent, and consequently superadd to the Commerce and profitable Export of much greater extent than the whole of her present traffic with the Coasts and Western Rivers of Africa". Veering perilously near hyperbole, Beaufoy compared the new route with Europe's discovery of the sea-passage round the Cape of Good hope in the fifteenth century — "another Geographical discovery less important indeed, but not less unexpected, now offers the means of effecting as complete a revolution in the Mercantile Intercourse with the Inland Countries of Africa".[60]

A combination of government caution, war with France, and unsatisfactory or unlucky explorers, kept some of the more grandiose schemes of the Association in the realm of vision rather than reality. It was Mungo Park and his great journey of 1795–7 which gave a new breath of life to the Association's activities. In his published account of 1799 Park remarked that his aim had been to "succeed in rendering the geography of Africa more familiar to my countrymen, and in opening to their ambition and industry new sources of wealth, and new channels of commerce".[61] Banks seized on Park's achievement to make his most telling address to the General Meeting of the African Association in May 1799: "We have already by Mr. Park's means opened a Gate into the Interior of Africa, into which it is easy for every Nation to enter and to extend its commerce and Discover from the West to the Eastern side of that immense continant". The gate, however, might not remain open to all and sundry. The British must return, not with single explorers, but with troops and field guns, establish forts on the banks of the Niger, and wrest control out of Muslim hands. Its gold production, he estimated, might be multipled a hundredfold, would draw in British manufactures, and open "the chance of an incalculable future encrease".[62]

This plea for annexation raised several issues, among them the rights of the existing inhabitants of the area. It was a few years earlier that one of Banks's correspondents had written to him from Sierra Leone, warning that the natives of that region "express on all occasions a consciousness that the soil, and country is their own, saying this is not white man's country, this belong to black man, who will not suffer white man to be master here, as he is in the East, and West Indies".[63] There was an answer, of a kind, to this; but first the deed must be done. Enclosing a forceful resolution from the General Meeting asking for the dispatch of troops, Banks wrote to his old friend, Hawkesbury, at the Board of Trade:

"Should the undertaking be fully resolved upon the first step of Government must be to secure to the British throne, either by Conquest or by Treaty the whole of the Coast of Africa from Arguin to Sierra Leone; or a least to procure the cession of the River Senegal, as that River will always afford an easy passage to any rival nation who means to molest the Countries on the banks of the Joliba [Niger]".[64]

Manufactured goods would flow in one direction, gold dust in another. The slave trade would be damaged, and the British trading corporation set up to do all this "would govern the Negroes far more mildly and made them far more happy than they are now under the tyranny of their arbitrary princes, would become popular at home by converting them to the Christian religion by inculcating in their rough minds the mild morality which is engrafted on the tenets of our faith". Christianity and commerce, conquest and treaty: with these recommendations to the government at the close of one century, Banks indicated much of what was to follow in Britain's dealings with Africa in the next.

The Banks of this paper is perhaps slightly different from the one so familiar to most of us: the patron of the sciences, cosmopolitan in outlook and friendships, ready to help scholars, collectors and travellers from all nations, a leading figure in an international fraternity of the learned. In many ways, especially in the diversity of his interests, Banks was exceptional; but he was also a man of his time. He belonged to a ruling class with firm notions of order and propriety; he lived in a country where rivalry with France had become endemic. For the most part Banks's attitude to individual Frenchmen was one of generosity and hospitality. He did much to strengthen the links between the learned societies of London and Paris, kept up cultural contacts between the two nations even in time of war, sped La Pérouse on his way with gifts which included navigational instruments used by Cook. But at times there was a sharper edge. In 1761, while still an undergraduate at Oxford, Banks joined his first society, the Society for the Encouragement of Arts, Commerce and Manufactures in Great Britain. Under this cumbersome name (usually, and sensibly, shortened to the Society of Arts) it sought to compete with France at all levels of economic and cultural activity, and this commitment long remained with Banks.[65] There was the suspicion of the French expressed in the *Endeavour* journal in April 1771 at Cape Town when Banks heard news of Bougainville's voyage.

"How necessary then will it be for us to publish an account of our voyage as soon as possible after our arrival if we mean that our countrey shall have the Honour of our Discoveries! Should the French have publishd an account of Mr De Bougainvilles voyage before that of the second Dolphin how infallibly will they claim the Discovery of Cypre or Otahite as their own, and treat the Dolphins having seen it as a fiction...if England chuses to exert her Prior Claim to it, as she may hereafter do, if the French settle it may be productive of very disagreeable consequences."[66]

There was prejudice, of an understandable nature in relation to the particular individual — but it goes beyond that — when in 1786 Banks's irritation at the behaviour of the French botanist L'Heritier spilled over into something more general: "Of all the impudent Frenchmen in the whole world, he is the most impertinent and dangerous".[67] In 1794 there was an expression of the realities of national rivalry in Banks's refusal to accept a would-be explorer from France in the service of the African Association (a curious suggestion anyway, since Britain and France were at war).

"Was a Frenchman to succeed even moderately in the investigation on which he might be employd are we not sure that his Countrymen would give the whole merit of the Adventure to the National Character of a Frenchman by the same rule as induced them to consider Bougainville who discovered nothing as superior to Cook who discoverd every thing — it is my opinion that they would do so and I fear that opinion is but too well founded in experience as well as in observation."[68]

For Banks there was no contradiction between a forthright pursuit of the material interests of his own country and promotion of the wider objectives of scientific investigation. He wrote, to a Frenchman as it happens, in 1788: "I certainly wish that my Country men should make discoveries of all kinds in preference to the inhabitants of other Kingdoms".[69] The founder members of the Society of Arts in 1754 would have recognised, and applauded, that sentiment.

Notes

Copies or transcripts of the documents cited in this paper from the Sutro Library, San Francisco, the Mitchell Library, Sydney, and the libraries of Yale University, Stockholm University, the University of London, and the Fitzwilliam Museum, Cambridge, are held at the Banks Archive, Natural History Museum, London. In writing this paper I have received much help from Harold Carter and Julia Bruce at the Banks Archive, as well as from Alan Frost, Dorothy Middleton and Andrew Cook.

(1) J.C. Beaglehole (ed.), *The Endeavour Journal of Joseph Banks 1768–1771*, Sydney, 1962, vol. 1, 108.

(2) H.B. Carter, *Sir Joseph Banks 1743–1820*, London, 1988, 169.

(3) James King to Banks, October 1780, Natural History Museum, London, Dawson Turner Copies [hereafter DTC], vol. 1, 304.

(4) See D. Mackay, *In the Wake of Cook: Exploration, Science & Empire 1780–1801*, London, 1985.

(5) James Matra to Banks, 28 July 1783, British Library [hereafter BL], Add.MS.33,977, fol.206.

(6) A. Frost, *Convicts and Empire: A Naval Question 1776–1811*, Melbourne, 1980, 10, 13–14.

(7) J. Dalrymple, *Memoirs of Great Britain and Ireland*, London, 1790, vol. 3, Appendix I, 313.

(8) Joseph Billings to Banks, 8 January 1783, Royal Botanic Gardens, Kew, Banks Correspondence [hereafter Kew, B.C.], vol. 1, 122.

(9) Sutro Library [hereafter SL], PN1: 3A, n.d.

(10) SL, 19A, 19B.

(11) William Hunter [to David Scott], 21 November 1786, SL, PN1: 11A.

(12) SL, PN1: 21, n.d.

(13) See G. Williams, *The British Search for the Northwest Passage in the Eighteenth Century*, London, 1962, chap. 10.

(14) See Mackay, op. cit.(4), 62–4.

(15) [J. Etches], *An Authentic Statement of All the Facts relative to Nootka Sound*, London, 1790, 2.

(16) N. Portlock, *A Voyage Round the World*, London, 1789, 6.

(17) Jonas Dryander to Banks, 19 September 1785, Fitzwilliam Museum, Cambridge.

(18) Richard Cadman Etches to Banks, 14 March 1785, Kew, B.C., vol. 1, 195; Banks to Carl Peter Thunberg, 17 June 1785, Stockholm University Library, Banks MSS, ALS.

(19) Banks to Henry Dundas, 15 June 1787, Yale University Library, Banks MSS.

(20) Etches to Banks, 17 July 1788, SL, PN1: 6. The context shows that Etches's meaning here is protection *by* government. This set of letters from Etches to Banks has been printed in F.W. Howay, 'Four letters from Richard Cadman Etches to Sir Joseph Banks, 1788–1792', *British Columbia Historical Quarterly* (1942) **6**, 125–39.

(21) Ibid.

(22) See M. Steven, *Trade, Tactics and Territory: Britain in the Pacific 1783–1823*, Melbourne, 1983, 47–9.

(23) Etches to Banks, 23 July 1788, SL, PN1: 7.

(24) Etches to Banks, 30 July 1788, SL, PN1: 8.

(25) SL, PN1: 1, n.d.

(26) See H. Fry, *Alexander Dalrymple and the Expansion of British Trade*, London, 1970, especially chap. 8.

(27) Dr Andrew Cook informs me that he has material on the Royal Society Club, hopefully soon to be published, which shows a much warmer relationship between Banks and Dalrymple than I suggest here.

(28) Alexander Dalrymple [to Evan Nepean?], 2 February 1790, Public Record Office [hereafter PRO], C.O.42/72, 501.

(29) Banks to Alexander Dalrymple, 30 December 1787, see W.R. Dawson (ed), *The Banks Letters*, London 1958, 250.

(30) On Wegg see G. Williams, 'The Hudson's Bay Company and its critics in the Eighteenth Century', *Transactions of the Royal Historical Society*, 5th Ser. (1970), **20**, 167–71.

(31) Alexander Henry to Banks, 25 March 1781, SL, PN1: 11.

(32) Henry to Banks, 18 October 1788, DTC, vol. 1, 39–51.

(33) George Dixon to Banks, 27 August 1784; Banks to Dixon, 29 August 1784, DTC, vol. 4, 47, 49.

(34) SL, PN1: 18C, n.d.

(35) J.M. Nooth to Banks, 4 November 1789, SL, PN1: 18A, printed in R.H. Dillon, 'Peter Pond and the overland route to Cook's Inlet', *Pacific Northwest Quarterly* (1951), **42**, 329.

(36) PRO, C.O.42/72, 495–8; SL, PN1: 18B.

(37) William Wyndham Grenville, 1st Baron, to Governor Arthur Phillip, March 1790 (draft), *Historical Records of Australia*, Sydney, 1914, ser. 1, vol. 1, 161–4.

(38) Dixon to Banks, 20 October 1789, SL, PN1: 2.

(39) Grenville to Commodore William Cornwallis, 31 March 1790, PRO, H.O.28/61, 273.

(40) SL, PN1: 17, and Mackay, op. cit.(4), 91.

(41) Grenville to Banks, 12 April 1790, PRO, H.O.42/16, 18.

(42) William Eden, 1st Baron Auckland, to Banks, 29 December 1790, DTC, vol. 7, 184–5.

(43) Nathaniel Portlock to Banks, 26 December 1790, Mitchell Library, MS.743/1.

(44) Mackay, op. cit.(4), 100. Dr Kaye Lamb has pointed out that "there is no evidence that Vancouver actually received" Banks's surveying instructions —"exercises in the obvious", W.K. Lamb (ed.), *The Voyage of George Vancouver 1791–1795*, London, 1984, vol. 1, 45, 46.

(45) Ibid, 102.

(46) Banks to Menzies, 22 February 1791, BL, Add.MS.33,979, fol.75v.

(47) H.B. Carter, *Sir Joseph Banks (1743–1820). A guide to the biographical and bibliographical sources*, Winchester, 1987, 65.

(48) For one particularly unpleasant example, see the 1789 letter of William Cadman Etches quoted in Steven, op. cit. (22), 43.

(49) Banks to Admiralty, 12 August 1815, University of London Library.

(50) R. Hallett (ed.), *Records of the African Association 1788–1831*, London, 1964, 44–5.

(51) R. Hallett, *The Penetration of Africa*, London, 1965, vol. 1, 212.

(52) For Banks's interest in Africa before the founding of the African Association see ibid., 175–6.

(53) Hallett, op. cit. (50), 74.

(54) Ibid., 102.

(55) Ibid., 117.

(56) Cited in H. Rutherford, 'Sir Joseph Banks and the Exploration of Africa, 1788 to 1820', PhD thesis, University of California, 1952, 23.

(57) Hallett, op. cit. (50), 118.

(58) Henry Beaufoy to Banks, 7 June 1796, SL, A2: 94.

(59) Hallett, op. cit. (50), 142.

(60) Ibid., 145–6.

(61) M. Park, *Travels in the Interior Districts of Africa*, 1799, 2.

(62) Hallett, op. cit. (50), 169.

(63) John Symmons to Banks, 30 March 1794, SL, A2: 76.

(64) Hallett, op. cit. (50), 272.

(65) For the Society of Arts, and for Francophobia generally in this period, see L. Colley, *Britons: Forging the Nation 1707–1837*, New Haven, 1992, 33–4, 88–91, 368.

(66) Beaglehole, op. cit. (1), vol. 2, 249.

(67) Carter, op. cit. (2), 225.

(68) Banks to Sir John [Stepney], 28 July 1794, SL, A2: 96.

(69) Banks to Buache de la Neuville, 20 November 1788, DTC, vol. 6, 89–90.

D.N. Robinson, 'Sir Joseph Banks and the Lincolnshire influence', in *Sir Joseph Banks: a global perspective* (eds. R.E.R. Banks and others), Royal Botanic Gardens, Kew, 1994, 193–195.

15. SIR JOSEPH BANKS AND THE LINCOLNSHIRE INFLUENCE

DAVID N. ROBINSON

Eastgate, Louth, Lincs

If the environment of childhood really does influence later life and actions, it was probably true of Joseph Banks. He was a two-year old toddler when the family returned from London to the large ancestral house in the deer park at Revesby on the edge of the Lincolnshire Fens. There he lived and explored for the next seven years, and then during some school holidays until he was nearly eighteen.

The Abbey, built about eighty years before and a mile from the ruins of the Cistercian abbey, was where his father, William, had grown up. But the same year the new family took up residence William was struck down with a mysterious fever which left him paraplegic and confined to his chair. Yet he maintained an indomitable good humour as he pursued his interest in the straightening of the River Witham and other drainage works, in planning gardens near the house and planting trees in the park.

So it was with his mother that Joseph first explored the gardens, the deer park and the village. It was she who told him when he discovered his first toad that it was a harmless animal. He enjoyed catching them and applying them to his nose and face, much to the consternation no doubt of his young sister but gaining the respect of the boys from the servants' hall and the village. It was probably with them he discovered the pleasures of swimming and fishing. He was always an active boy and learned to ride a horse at an early age.

When Joseph was about seven his father engaged Revd Henry Shepherd, rector of Mareham le Fen and Moorby, as a private tutor, as he had been to William twenty years before. The presentation to those livings was in the gift of the Bishop of Carlisle who was lord of the manor of nearby Horncastle where William was lessee of his lands. Probably Joseph travelled in the coach to Horncastle with his father when he went there on business. We do not know whether or not he met the Bishop, Dr Charles Lyttleton, who was a noted antiquarian (and was among Joseph's supporters when he was later nominated as a Fellow of the Royal Society). However, one can easily imagine young Joseph taking an interest in the substantial remains of the walls of the Roman military fort in Horncastle.

At the age of nine he was packed off to Harrow where he missed the broad acres of Lincolnshire, and doubtless revelled in the late summer months of 1756 at Revesby before being moved to Eton School. Now big and strong for his age, he made friends easily and was old enough to ask questions of his father about drainage works when fellow promoters came to confer with William at the Abbey. Perhaps he was taken to see the adjoining estate woodlands at Tumby and Fulsby where there had been recent felling and replanting under the

23-year rotation management. Or a chance to see the Wildmore Tit ponies rounded up from Wildmore Fen, and other strings of horses on their way to the Great August Horse Fair at Horncastle which was developing into the largest horse fair in the world. The as yet undrained and still intercommoned fens of East, West and Wildmore, covering some 40,000 acres, came almost to the gates of the deer park at Revesby. Joseph's father had an interest there through his own rights and those of his tenants to graze sheep, cattle, geese and horses, and he would have grown up with a knowledge of their part in the farming economy of the area.

But it was the lakes and meres of the East Fen which fascinated Joseph. The result of medieval peat digging, they formed a chain of shallow interconnected waters fringed with reed. A boat could be taken along Cut Syke Row, into Stickford Syke and through Steven Water into Silver Pit, Arch Booze and Goop Hole. The clear water abounded in fish which could be speared with a hayfork. And beyond the meres lay some three hundred acres of mossbery or cranberry fen. Not only was the East Fen good for fishing but it was also rich in birds and butterflies and a botanical paradise with some 200 species. This was to interest Joseph much more after the incident at Eton when he determined to know the names of hedgerow plants and engaged the local cullers of simples to teach him about them. More and more time out of school was spent armed with his mother's battered copy of Gerard's *Herball* hunting plants and insects and collecting shells and fossils.

By the time he was a lanky lad of seventeen (and nicknamed "Shanks"), Joseph had already established a well organised herbarium and other collections. In July 1760 he was brought home to Lincolnshire for a course of innoculation for smallpox. It did not take and a second course was necessary; by the time he had recovered it was not worth returning to school. So for five months, before he entered Christ Church, Oxford, he was free to pursue his countryside interests with a degree of dedication and determination, a period of pre-academic fieldwork. Time to explore the heathlands beyond Tumby and Tattershall across the River Bain, and to pursue some antiquarian interest at the keep of Ralph Lord Cromwell's castle at Tattershall and the ruins of his hunting lodge on the open moors. Maybe he also explored into Wildmore Fen, the "willow mere" where later after drainage a spot would be known as Botany Bay. Certainly he took a boat again through the East Fen meres where he became more familiar with the distinctly coloured and textured natterjack toad (the local name being the "natter jack" because of the churring nature of its call), which he was later to describe as a new British species. More often than not Joseph was on his own rather than taking a servant when botanising round the farms and quarries in the sandstone Wolds, and it was not unknown for him to be mistaken for a vagabond and ordered off the property.

He would not see Lincolnshire regularly again for nearly twenty years, although he continued to take a close interest in estate and county affairs after his father's death and inheritance of the estates. In November 1761 he donated £23 towards the preliminary expenses of obtaining the act of parliament for improving the Witham, the scheme so dear to his father's heart. And in 1775 a nine-bay house in High Street in Horncastle was completed which was to become the Banks town house (and still stands). Following his marriage in 1779, Banks visited his principal country seat at Revesby every year until within three years of his death. Many county affairs had to wait until that autumn visit, so vital was his advice, opinion and preferably presence at meetings. As Lincolnshire had influenced young Joseph, so in later years did Sir Joseph influence affairs in the county

Among them were the production of wool, the establishment of a woollen mill at Louth and promotion of the annual county Stuff Ball; the development of turnpikes and the construction of canals particularly that to Horncastle; the Lincolnshire Medical Benevolent Society and the establishment of the Horncastle Dispensary. The connection with Horncastle has always been strong and a street in the town is named after him. He served as Lieutenant Colonel of the Lindsey Battalion of the Supplemental Militia when there were fears of enemy landings on the coast, he served his year as High Sheriff of the county, and was Recorder of Boston. He took an interest in geological matters, visiting the submerged forest at Huttoft, kept in touch with archaeological finds, employed Claude Nattes to make sketches of over seven hundred churches, country houses and monuments in the county, and helped to forward the first ordnance survey of Lincolnshire. It could be said that Banks had the nucleus of a county museum including archives and portraits at Revesby Abbey.

When Arthur Young of the Board of Agriculture visited Revesby Abbey he testified to the orderly nature of the filing system in Banks's offices. On the same occasion Banks took him in a boat through the East Fen meres with which he was so familiar. Yet when engineer John Rennie came up with a scheme three years later (1800) to drain the last of the Lincolnshire fens, Banks was the active driving force behind it. The open field farming of parishes surrounding Revesby, many with Banks as lord of the manor, had been enclosed and improved by acts of parliament, the Witham from Chapel Hill to Boston had been straightened in 1766, but it took the famine years at the close of the century and the need to increase food production in the Napoleonic wars to trigger the final drainage. The enormous and costly scheme (over £600,000) took only thirteen years to complete.

While accepting the inevitable outcome of the application of science to land improvement, Banks seems to have regretted the disappearance of those pellucid pools of the East Fen, beloved of his youth. In a poem written in 1807 when the drains were being dug, he lamented the demise of "Gowple's filthy meer" and "Eweholme's Reeds by Starlings broke", and "all those pretty little tracts Where Pike now prey on Eels and Eels on Jacks".

When bread prices were high in 1800–01, Banks shipped rice to Boston for sale cheaply with herring to his tenants, and supplied a recipe for cooking them. When he saw a tenant improving his land by hollow drainage he would give a lease of 21 years as a reward and an encouragement. On the occasion of the annual Revesby Fair in October, a bullock was killed, the strongest ale in the Abbey cellars was broached, the servants' hall thrown open for the day, and beef and beer was distributed to all. And if he did not see enough of his people jolly he said that the ale had not been brewed strong enough. The tradition of serving beef and beer at the Revesby estate annual tenants' lunch continues to this day.

E.R. Andrew, 'Sir Joseph Banks and Boston', in *Sir Joseph Banks: a global perspective* (eds. R.E.R. Banks and others), Royal Botanic Gardens, Kew, 1994, 197–200.

16. SIR JOSEPH BANKS AND BOSTON

E. RAYMOND ANDREW

Department of Physics, University of Florida, Gainesville, Florida, U.S.A.

This poster paper is concerned with Sir Joseph Banks and his activities in Lincolnshire where he owned family estates at Revesby Abbey, near Boston, inherited from his father at the age of 18. His great-grandfather Joseph Banks (the first) purchased the Revesby estates in 1714 and Sir Joseph Banks (the fourth) succeeded to the estates in 1761. Despite his very active life as an explorer and man of science and affairs in London, Sir Joseph retained a detailed practical interest in his Revesby estates all his life. He visited Boston and Revesby several times a year including an annual migration with his wife and his sister from August to October.

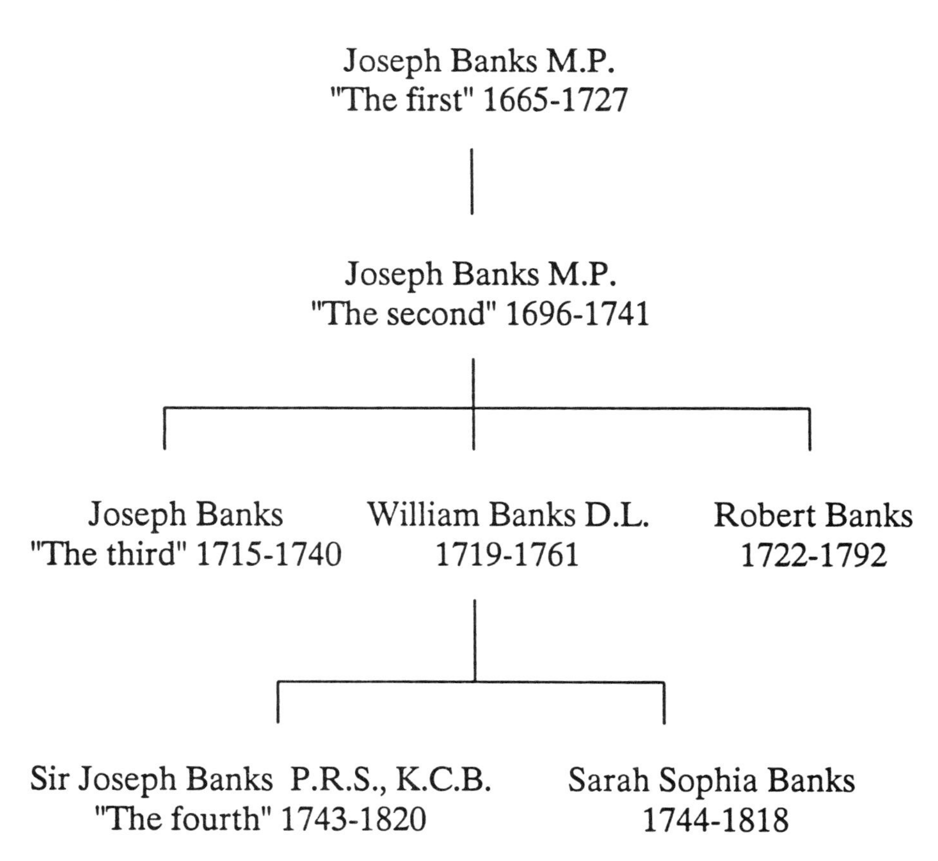

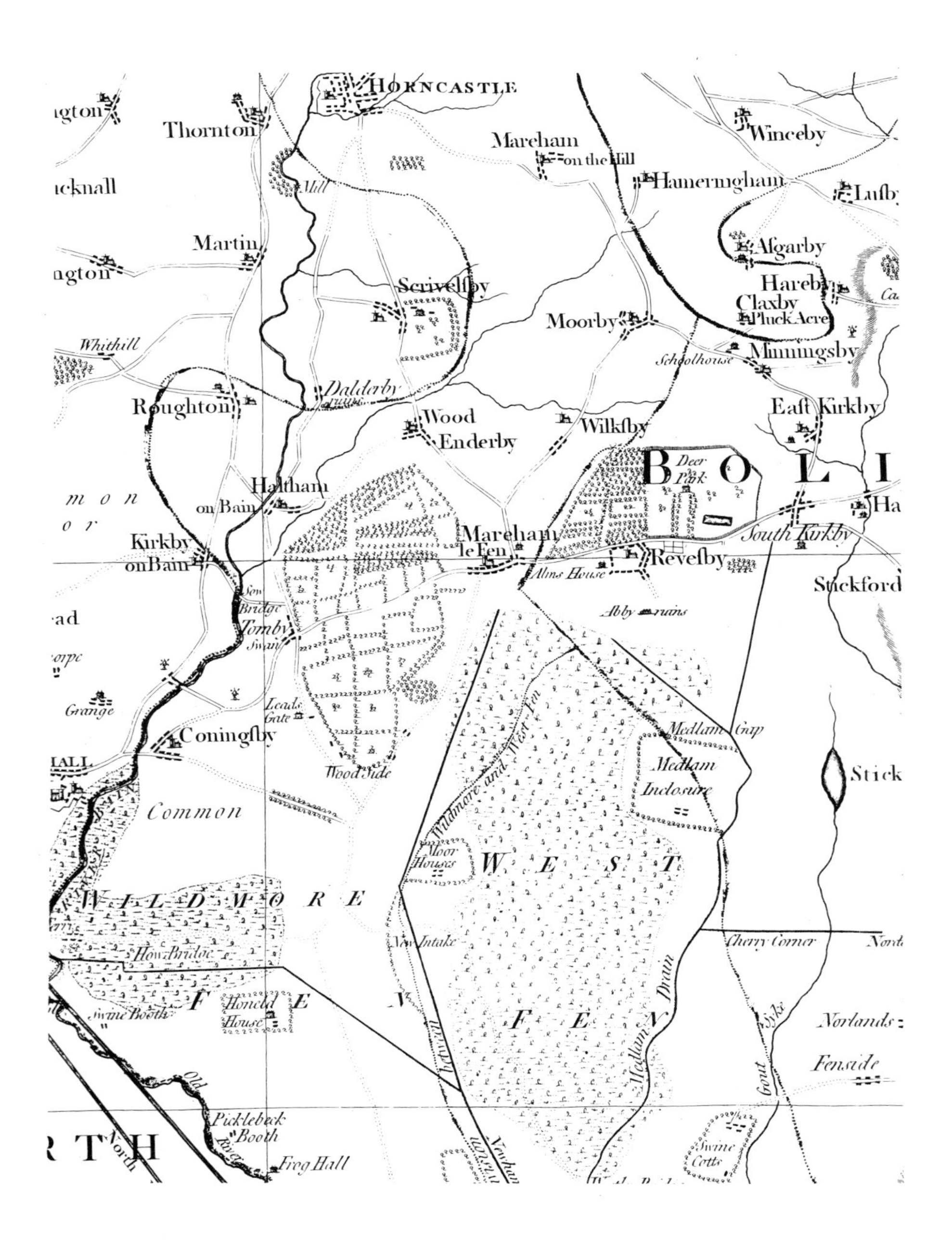
HORNCASTLE
Thornton
Mareham on the Hill
Wineby
Hamerngham
Lusby
Martin
Asgarby
Scrivelsby
Hareby
Claxby
Pluck Acre
Moorby
Whithill
Schoolhouse
Minningsby
Dalderby ruins
Roughton
Wood Enderby
Wilksby
East Kirkby
B O L I
Deer Park
Haltham on Bain
Kirkby on Bain
Mareham le Fen
South Kirkby
Revesby
Alms House
Stickford
Abby ruins
Tomby
Swan
Leads Gate
Coningsby
Medlam Gap
Wood Side
Medlam Inclosure
Common
Wildmore and West Fen
Moor Houses
W E S T
W I L D M O R E
New Intake
Cherry Corner
How Bridge
Swine Booth
F E N
Honeld House
Norlands
Fenside
F E N
Medlam Drain
Gout Syke
Pickleback Booth
Frog Hall
Swine Cotts
Old River
R T H

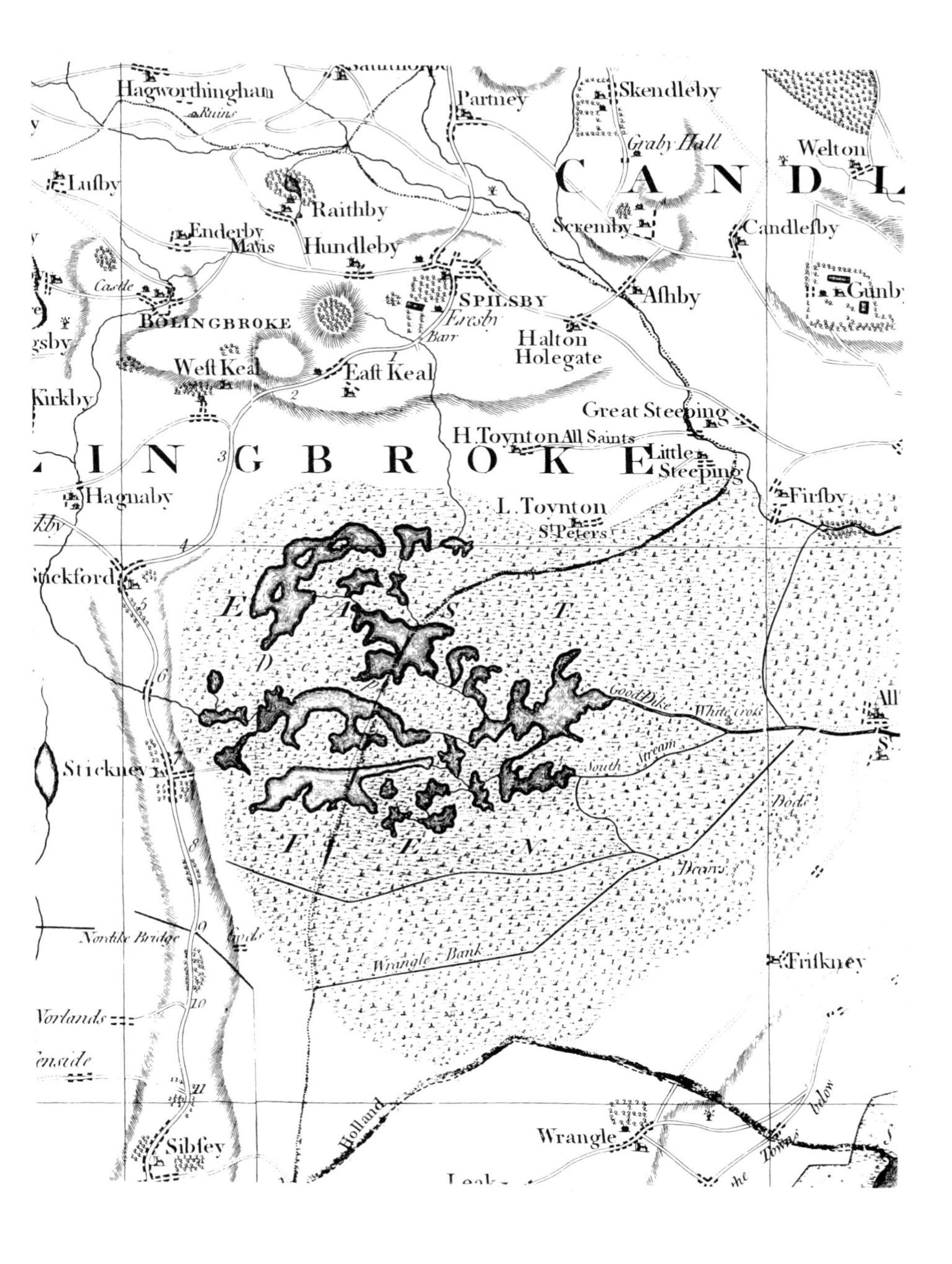

Hagworthingham
Ruins
Partney
Skendleby
Graby Hall
Welton
C A N D L
Lufby
Raithby
Enderby
Mavis
Hundleby
Screnby
Candlefby
Castle
BOLINGBROKE
SPILSBY
Erefby
Barr
Afhby
Gunby
Halton
Holegate
Weft Keal
Eaft Keal
Kirkby
Great Steeping
H. Toynton All Saints
Little
Steeping
I N G B R O K E
Hagnaby
Firfby
L. Toynton
St Peters
Stickford
E A S T
Deeps
Good Dike
White Crofs
South Stream
Stickney
F E N
Decoys
Nordike Bridge
Wrangle Bank
Frifkney
Holland
Wrangle
Sibfey

His early interest in botany was enhanced in his school holidays in Revesby with the aid of Gerard's "Herbal" found in his mother's dressing room there. The Revesby estate furnished two servants, Peter Briscoe and James Roberts, to accompany him with Captain Cook to the South Seas in the Endeavour. Their names along with that of Joseph Banks, Squire of Revesby, are listed among explorers of Australia perpetuated on a table in Boston Parish Church.

Sir Joseph Banks was a model Lincolnshire landowner and estate manager. He regarded agriculture as an application of natural science and his authorative views on crops and stock were in demand. He had a lifelong interest in wool and was an expert on merino sheep and wool. He kept several hundred deer in the park at Revesby. He computed the economics of his deer-keeping and found he usually made a profit from it. He was very concerned with the state of the fens which at the time were in a most neglected state, continuously flooded and overgrown. He took the initiative and chaired meetings of fenland proprietors in Boston Town Hall to agree upon a drainage plan under Sir John Rennie.

Other local interests in which Banks played a part included improvement of the navigation of the river Witham, renovation and deepening of Boston harbour, and construction of the Horncastle Canal. He participated in the boring for water in central Boston and communicated a paper to the Royal Society on the strata found there.

Sir Joseph was an authority on archaeology in Lincolnshire and his reputation was such that any discovery was reported to him at once. He had a wide knowledge of Lincolnshire churches and commissioned four volumes of drawings of them now in Lincoln City Library. They are valuable in showing the condition of the churches before later restoration and alteration. He opposed the removal of the twin spires from the western towers of Lincoln Cathedral in 1807, but found the Dean and Chapter were not to be diverted from this enterprise.

From 1809 Sir Joseph was Recorder of Boston and he took his turn as High Sheriff of Lincolnshire. He was following a family tradition of local public service; his great grandfather had been M.P. for Grimsby, his grandfather M.P. for Peterborough and his father Deputy Lieutenant of Lincolnshire. In view of his wider activities in London and the world at large, the degree of his energetic interest and beneficial participation in local affairs in Lincolnshire is truly remarkable.

S. Bell, 'Sir Joseph Banks and Aimé Bonpland', in *Sir Joseph Banks: a global perspective* (eds. R.E.R. Banks and others), Royal Botanic Gardens, Kew, 1994, 201–204.

17. SIR JOSEPH BANKS AND AIMÉ BONPLAND

STEPHEN BELL

Department of Geography, McGill University, Montreal, QC, Canada

The French naturalist Aimé Bonpland (1773–1858) usually has his name linked with Alexander von Humboldt (1769–1859), on account of the great scientific journey undertaken together in equinoctial America from 1799–1804. After an interlude of mainly botanical work in France, he resided from late 1816 until his death in various parts of southern South America, where he combined a wide range of occupations, including the practice of medicine, pharmacy, plant collecting and merino breeding. In contrast to Humboldt, Bonpland is almost a forgotten figure today. His links with England and with Sir Joseph Banks have not attracted much research attention.

As well as vast differences of character and circumstance, there are interesting parallels between Banks and Bonpland. Both were involved in assembling major herbaria, both wrote a lot but published little, both served as clearing-houses for scientific information, albeit on very different scales. Just as Banks served the Crown in Britain through his efforts to develop merino breeding, Bonpland held responsibility for the flocks kept by Joséphine of France. Given these common interests, it is worth examining how a scientific relationship developed, a task constrained by a fragmented historical record. The following represents an early stage of the author's research on Bonpland, stimulated by other work in South American archives.

From Equinoctial America to Malmaison

Bonpland came to Banks's attention through Humboldt, who first visited England in 1790. Six years later, Humboldt inherited the considerable means that allowed him to plan a major scientific expedition. Turning to Banks for help in connection with a proposed trip to the Levant, in 1798 Humboldt asked for a letter of support clearing his motives. In this bold gesture, he asked that the document should also name Bonpland, the friend who was going to help with the herborization.

After a series of false starts, Humboldt and Bonpland went on their long trip to Spanish America (Figure 1), where the latter's contribution proved critical, especially in the field of botany. Most of the 5,800 species of plants collected during the American journey, the greater part of them new to science, resulted from Bonpland's efforts. While a capable correspondent, Bonpland could not match Humboldt in keeping the scientific world informed of their achievements. His zeal for searching out new plant species possibly left little time for writing. In

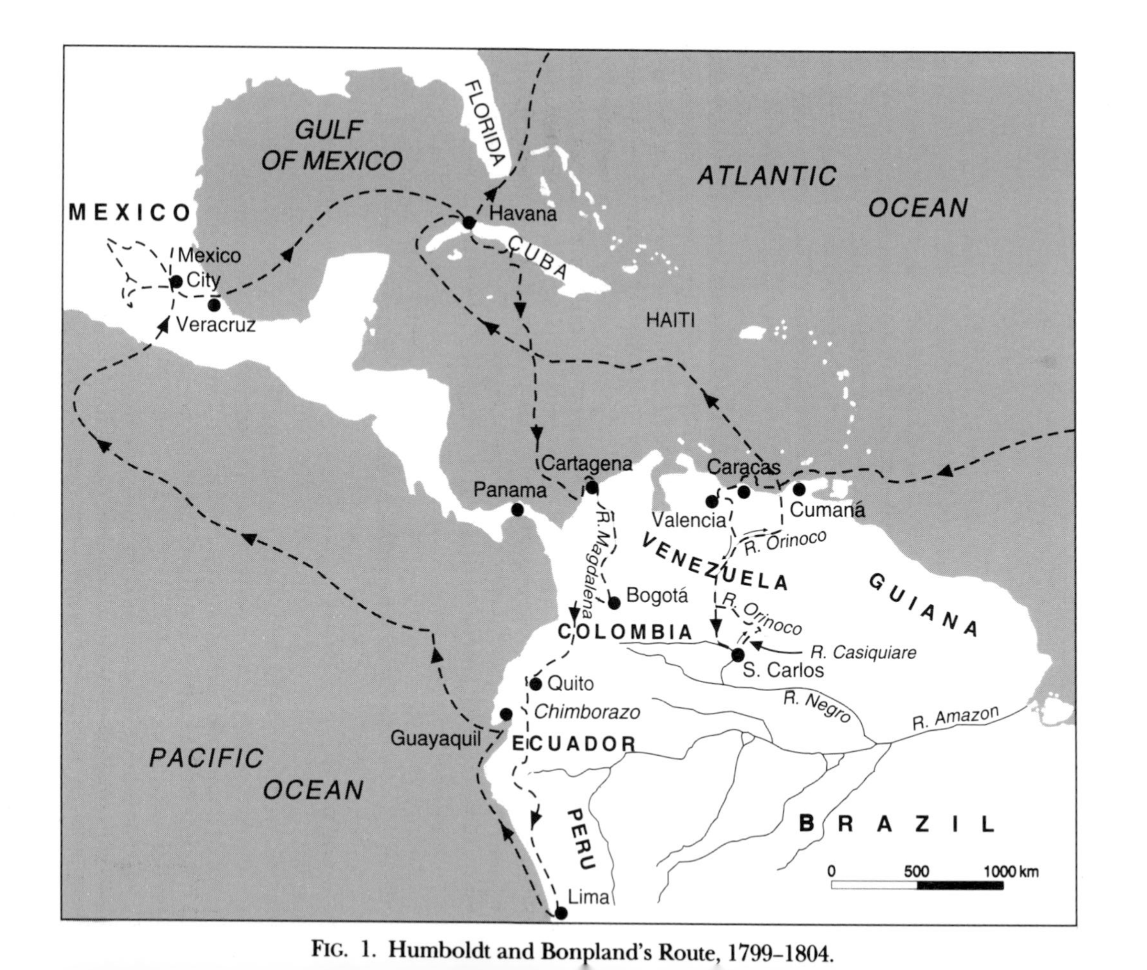

FIG. 1. Humboldt and Bonpland's Route, 1799–1804.

Humboldt's letter to Banks from Cumaná, Venezuela in November 1800, Bonpland, "pupil of Jussieu and Desfontaines", received merely a brief commendation for his energy and knowledge.

Circumventing the obstacles of politics and pirates, Humboldt took great pains to maintain contacts with the leading botanical establishments of Europe. By the time he reached Mexico in 1803, he had already dispatched twelve lots of seeds to different recipients, including to Kew and to Banks. No doubt in the minds of seed recipients, Bonpland also received some of the credit for this extraordinary activity.

On return to Europe in 1804, the lives of the travellers soon began to diverge as Humboldt broadened his circle of scientific collaborators. During the next decade, Banks continued to hear much more from and of Humboldt than Bonpland, yet he was also made aware by Sir Charles Blagden that the latter was playing a major role in preparing the botanical publications. The relationship between Banks and Bonpland moved into a new phase in 1809 with a letter from the former. It seems probable that Banks would have been interested in Bonpland's recent appointments as the botanist and intendant at Malmaison. When Bonpland took up his duties at this important establishment, it already included active research missions in plant descriptions and experimental cultivation, which counted on Banks's support.

This period in Joséphine's employ (1809–14) was undoubtedly a high point of Bonpland's career. Earlier and later very much in the shadow of Humboldt, this fruitful spell of work with few financial constraints provided an opportunity to develop his own scientific profile, encouraged by Joséphine. It is during this time that Bonpland pursued a link with Banks. On 8 February 1810, Bonpland wrote a matter-of-fact letter to him, asking to be placed in contact with a bookseller, as well as an agent to supply plants, especially seeds. Within two weeks, he was writing again, sending the first three parts of the second volume of *Plantae aequinoctiales*. These were carried to London by Madame Solvyns, an Englishwoman by origin. For plants and seeds, Bonpland told Banks that he had written to the Scottish botanist John Fraser, whom he knew from Havana on the journey with Humboldt. In addition, he wanted to be placed in contact with Lee and Kennedy's famous nursery, The Vineyard at Hammersmith, known to him only by reputation. Finally, by writing, Bonpland hoped to attact some more interesting seeds for his work, claiming that he would sow and cultivate these personally.

Bonpland in England

While at Malmaison, Bonpland managed to travel widely for his work in applied botany, to Berlin and Vienna for example. Although England was closed to him on account of war, his letters of 1810 to Banks reveal clearly his latent interest in British botanical developments. With Napoleon's resignation in 1814, Bonpland became an early representative of a stream of scientific visitors from France, first crossing to England in July. His brief description of this 1814 journey supplements the fuller and published account of a trip to England in this period by his colleague Augustin-Pyramus de Candolle (1778–1841). With Banks at their centre, Bonpland and de Candolle traced much common ground in their scientific contacts in and around London.

One of the first things Bonpland did was to visit gardens containing rare plants, such as the vast establishment of Lee and Kennedy. He saw Kew in the company of Sir Charles Blagden, Robert Brown and William Aiton the younger. At Kew, Bonpland dined with Banks, who showed him his exotic trees, in particular araucaria pines from Peru. Bonpland also frequented Banks's herbarium in Soho Square, where he made a list of the Melastomaceae, the genus he was working on at the time. Back in London again in early 1816, he made further descriptions based on Banks's herbarium, notably of cinchona, which received an extensive description in the first volume of *Plantae aequinoctiales*, finished in 1817. He also described the begonia using Banks's specimens. The many transcription errors in Bonpland's manuscripts indicate that he had little English but this may not have been an insurmountable obstacle in communicating with Banks, as de Candolle's recollections show.

Bonpland's intense traffic across the Channel was motivated only in part by the rich resources placed at his disposal by Banks. Much later, reflecting from South America on the French portion of his career, Bonpland maintained that he had sought to use his position in France during 1805–14 to help the independence movements of the South American countries. Much of the rationale for his multiple trips to England was to develop a closer relationship with Simón Bolívar, through his London agents, and to be of greater use to the American cause.

Joséphine's death in 1814 robbed Bonpland of a protector. He was soon giving serious consideration to the idea of quitting Europe for South America. As early as the autumn of 1814, Bonpland was preparing his way to Buenos Aires by sending seeds from Paris, some of them perhaps given to him in England. Most of these seeds were addressed to powerful political figures. During his time in London, Bonpland became particularly known to and friendly with Manuel Belgrano, Manuel de Sarratea and Bernardino Rivadavia, influential figures in the politics at Buenos Aires and leading conspirators for the independence of Argentina from Spain. Between them, these Argentines essentially recruited Bonpland. At the age of 43, he embarked in late 1816 for Buenos Aires, loaded up with a vast baggage of European plants. He spent more than four decades in southern South America, never returning to Europe.

Although Banks and Bonpland shared some common interests, their direct contacts were fleeting, coming as they did near the end of Banks's career. It is clear from Bonpland's plant descriptions that Banks made an important contribution to the botanical development in Malmaison. Also open for examination is Banks's influence in Bonpland's exchanges of seeds between the Old and New Worlds. The material in this communcation offers glimpses in the web of scientific communication. It is only one fragment of a large and enduring theme of Banks's late career, that of keeping the lines of communication open between Britain and France during decades of war. As Cuvier emphasized in a eulogy on Banks pronounced in Paris in 1821, "when the seas were closed in upon us on all sides, his voice opened a passage to scientific expeditions. Geography and Natural History are indebted to his solicitude for the preservation of their most valuable labours."

S. Minter, 'Sir Joseph Banks and the Chelsea Physic Garden', in *Sir Joseph Banks: a global perspective* (eds. R.E.R. Banks and others), Royal Botanic Gardens, Kew, 1994, 205–208.

18. SIR JOSEPH BANKS AND THE CHELSEA PHYSIC GARDEN

SUE MINTER

Chelsea Physic Garden, London

Summary

The poster was primarily given to air the work of Mark Laird while Historical Research Fellow at the Chelsea Physic Garden under support provided by the British Petroleum Company p.l.c. "Sir Joseph Banks Botanist, Horticulturist and Plant Collector: Associations with Chelsea Physic Garden". This was published to coincide with the inaugural planting of the Banks Display by HRH the Duke of Kent on 24 June, 1988, and in conjunction with the Australian bicentennial celebrations.

The poster concentrated on two aspect of Banks's contributions to the Physic Garden; firstly his introduction of species as a plant collector and patron of horticulture and secondly his donation of Icelandic lava in 1772. From this one of the earliest rock gardens in Britain was built in the summer of 1773 and remains to this day as a "listed structure". A map of the Garden identifying its location was presented, as well as the earliest known illustration of it (an etching by William Burgess of 1896) and colour photographs of the feature as it is today.

Plant collector and patron of horticulture

In 1761 Banks's father died, leaving him at 21 (1764) heir to estates worth £6,000 per annum. His mother moved to Turret House, Paradise Row, near Chelsea Physic Garden and Banks made the acquaintance of its gardener Philip Miller, who introduced him to the Garden's exotic collections, and of a circle of naturalists and nurserymen. These included (by 1765): Daniel Solander, John Lightfoot, James Lee, Thomas Pennant, William Watson, and the Duchess of Portland.

Philip Miller became a mentor at Chelsea Physic Garden. He no doubt stimulated Banks to thoughts of plant-collecting in distant lands, and fostered in him a life-long passion for exotic floras.

In return, Banks was to leave his mark at Chelsea: firstly, by donating seeds collected by himself or his agents across the world; and secondly, by bestowing on the Garden an unusual gift of Icelandic lava. Banks's enduring horticultural legacy is also evident in the New Zealand "Kowhai", *Sophora microphylla*, growing today against the Curator's house: this is considered, by tradition, to be a direct descendant of the plant, first introduced to cultivation in the Garden by Banks in 1772, from which William Aiton described the species in 1789.

The Banks Display in the present day Garden's "Historical Walk" represents geographical zones from New Zealand and Australia to China and the Caucasus. It celebrates the Garden's link with Banks through plants: a) named after him, notably the genus *Banksia*; b) for which he remains either the botanical authority or the basionym authority; c) first introduced by him to cultivation in Great Britain, whether directly as a result of his voyages or later, through his patronage.

If Philip Miller provided the initial foretaste of exotic floras at Chelsea Physic Garden, Joseph Banks's own practical experience of plant collecting began in Newfoundland and Labrador in 1766. He gathered some 340 species including *Kalmia polifolia* Wangenh. and *Rhododendron canadense* (L.) Torr. Both were new to cultivation in Britain. Interestingly, they were growing at Chelsea Physic Garden by the 1770s. *Rhododendron canadense* is recorded in the catalogue of 1772/3 as planted in the newly-made "swamp" garden. The kalmia, then known as "K. glauca", was amongst the plants depicted by Mrs Delany in 1779, which she had obtained from the Garden at Chelsea.

Banks's voyage to the South Pacific with Solander on the "Endeavour" (1768–1771) was extremely successful, some 1400 species in total being estimated new to science. The introduction of the first Australian and New Zealand plants to horticulture through Kennedy and Lee's nursery and Chelsea Physic Garden forms an important chapter in the history of gardening in Great Britain. However, the role Banks played in the first botanical description of such exotic species is complex on account of his failure to publish. It was left to other botanists to make use of the work accomplished by Banks and Solander, acknowledging their debt through the published botanical name. As a consequence, Banks and Solander, or sometimes either Banks or Solander individually, remain the authority or basionym authority for a range of plants, predominantly from Australia and New Zealand, which were described in publications from Josef Gaertner's *De Fructibus et Seminibus Plantarum* (1788/90) to Robert Brown's *Prodromus Florae Novae Hollandiae* ... (1810). In other cases, whilst the name used by Banks and Solander was not adopted, their herbarium specimens provided the basis for the description. Robert Brown, for example, indicates this through the abbreviation "B.v.s.", or Banks *vidi siccam*, attached to some 100 Australian species. Alternatively, the acknowledgement to Banks is implicit in the adoption of the specific epithet *banksii* for a group of plants such as *Grevillea banksii* R.Br. or *Cordyline banksii* Hook.f. The genus *Banksia* is the crowning symbol, acquiring its name from the younger Linnaeus in 1781.

Despite Kew's ascendency after 1773, Banks continued to ensure that exotic seeds reached Chelsea Physic Garden in 1781, 1789 and 1792; in return, specimens were sent from the Garden to Banks on a regular basis until 1792. This represented a period when Banks promoted plant collecting beyond Australia and New Zealand into China, the East Indies, South Africa, North America and Latin America, Russia and Asia Minor. In the minutes of the Garden Committee to the Society of Apothecaries we may read, for example, the entry for 21 May 1781: "The Gardiner informed the Committee that he had received from Sir Joseph Banks more than five Hundred different Kinds of Seeds of Plants which were collected in the late Voyage round the Globe and also more than one Hundred different Kinds of Seeds of Plants from Mr Alexander Anderson from Saint Lucia". Of the 449 species, whose introduction Miller's pupil and later, first royal gardener at Kew W.T. Aiton attributes directly to Banks, a representative selection take their place in the present day Banks Display at Chelsea Physic Garden and are listed below.

a) Plants collected from New Zealand and Australia during the *Endeavour* voyage 1768–1771

i) Those introduced to cultivation in Britain in 1771 (14 plants):
e.g. Haloragis erecta, Leptospermum scoparium, Sophora microphylla and Tetragonia tetragonioides from New Zealand.
Dianella caerulea and Eucalyptus gummifera from Australia.

ii) Those named subsequently from herbarium specimens (over 50 plants):
e.g. Astelia banksii, Pittosporum crassifolium and Senecio banksii from New Zealand.
Alyxia spp., Banksia spp., Callistemon viminalis, Melaleuca nodosa and Pittosporum spp. from Australia.

b) Plants introduced to cultivation in Britain from New Zealand and Australia through later collectors working with Banks

i) Those introduced from New Zealand (2 plants):
Acaena anserinifolia (1796) and Phormium tenax (c. 1789).

ii) Those introduced from Australia (c. 90 plants):
e.g. Acacia verticillata (1780), Araucaria heterophylla (1793), Callistemon citrinus (c. 1788), Calomeria amaranthoides (1800), Cissus antarctica (1790), Hardenbergia violacea (1790), Kennedia rubicunda (1788), Lagunaria patersonii (1792), Leptospermum flavescens (1787), Pittosporum revolutum (1795).

c) Plants introduced to cultivation in Britain from the rest of the world

i) By Banks himself:
From the *Niger* voyage to Newfoundland and Labrador (1767):
Kalmia polifolia and Rhododendron canadense.
From the *Sir Lawrence* voyage to Iceland (1772): Koenigia islandica and Salix myrtilloides.

ii) By other collectors working with him (over 300 plants):
e.g. Diospyros kaki (1789) and Eriobotrya japonica (1787) from Japan; Chaenomeles speciosa (1796), Hydrangea "Sir Joseph Banks" (c. 1788), Ligustrum lucidum (c. 1794), Magnolia denudata (1789), Paeonia suffruticosa (1787) and Rosa banksiae (1807) from China; Strelitzia reginae (1773) from South Africa; Campanula alliariifolia (1803), Centaurea dealbata (1804) and C. macrocephala (1805) from Mount Caucasus; Pinus banksiana (before 1785) and Zizania aquatica (1790) from North America; Penstemon barbatus and P. campanulatus (1794) from Mexico.

The rock garden

In 1772, preparations for a second voyage to the South Pacific on *Resolution* were abandoned by Banks in May after disagreement with the Admiralty over

Etching by Walter Burgess, 1896. This is the earliest known illustration of the 1773 rock garden made of Banks lava (to the left of the statue). The original is held at the Chelsea Physic Garden.

the ship. Banks and Solander sailed in the *Sir Lawrence* to Iceland on 12 July, returning in November with the lava as ballast. In 1773 the lava was made into "rock work" at Chelsea Physic Garden, May–August.

The minutes of the Garden Committee for the Society of Apothecaries record that on 16 April 1773 it was "Agreed that a piece of artificial Rock work will be a very ornamental addition to the Societys Garden as also very useful for the cultivation of such plants as will only thrive in stoney soil". It was to be composed of three types of material: Icelandic lava, traditionally regarded as from Mount Hekla, but now analysed as ballast taken on board the Sir Lawrence elsewhere in Iceland, 40 tons of masonry donated by Stanesby Alchorne from the Tower of London; and flints and chalk donated by Mr Chandler. By 16 August William Curtis and his assistant had completed the construction. There is sadly no direct evidence of what was being grown on the "rock work". The alpine *Primula marginata* Curt., introduced in 1781 and named by the creator of the rock garden, would perhaps have been included. It represents a group of plants being introduced from the Alps at this time through the influence of Dr John Fothergill and Dr Pitcairn. As they repeatedly contributed seeds to the Garden between 1771 and 1773, it is possible they provided seeds sent back by Blakie after 1775.

M. Fitton & S. Shute, 'Sir Joseph Banks's collection of insects', in *Sir Joseph Banks: a global perspective* (eds. R.E.R. Banks and others), Royal Botanic Gardens, Kew, 1994, 209–211.

19. SIR JOSEPH BANKS'S COLLECTION OF INSECTS

MIKE FITTON and SHARON SHUTE

Collections Management Division, Department of Entomology, The Natural History Museum, London

Banks's collection of insects is of great scientific importance and historical significance. It was given originally to the Linnean Society and later presented to the British Museum. The collection includes insects captured by Banks during Cook's circumnavigation of the globe. Not only were these specimens some of the first to reach England from areas such as Australia, but in London they were soon studied by the entomologist J.C. Fabricius. He described many new species from Banks's material and hence the specimens are taxonomic "types". Surprisingly, there is no comprehensive record of the collection. Much of the material now needs conservation and the cabinets housing the collection have deteriorated to a point where there is a pressing need to rehouse it. Proposals for support for essential conservation work, to catalogue the collection and to research its history are being formulated. Such a project would also throw further light on Banks's entomological interests.

Joseph Banks's collection insects is housed in The Natural History Museum, London. It is of continuing scientific importance and historical significance and is one of the few general insect collections maintained as separate entities in the museum. The collection was originally given to the Linnean Society, but in 1863 the society decided that upkeep of a general museum of zoology and botany was beyond its means. The society resolved "That the Collection of Insects and Shells of the late Sir Joseph Banks be presented to the British Museum". Apart from the insects and shells no other Banks material was involved and most of the society's other collections went elsewhere or were sold. When it arrived at the British Museum the collection was still housed in Banks's original cabinet. However, the drawers were considered unsuitable and not dust proof and the specimens were transferred to "better housing". Banks's cabinet went first to the Victoria and Albert Museum and subsequently to Cambridge, where it has recently been restored.

It is said that Banks began collecting insects during his school days and some of those specimens may survive among the British material in the collection. Like Banks's other collections, the insect collection contains material from many parts of the world, much of it collected on the voyages of some of the most famous 18th century explorers, such as James Cook, Matthew Flinders and William Bligh. Banks himself visited, and brought back collections from, Newfoundland and Labrador in 1766. Two years later, in 1768, he set sail with Cook in the Endeavour, on its circumnavigation of the globe, and during the voyage collected in many localities, including Madeira, Rio, Tahiti, New Zealand, Australia, New Guinea, Suva, Batavia, St Helena and parts of Africa. These were the some of the first insects to reach England from these regions. The material from Australia and New Zealand is of particular importance, constituting samples of the faunas before they were influenced by Europeans. After the return of the Endeavour in 1771, Banks made only one further expedition: to Iceland and the Shetlands. However, he sent and encouraged others to collect for him. For example, David Nelson accompanied Cook on his third voyage, in the Discovery.

Johann Fabricius, a former student of Linnaeus, visited London in the early 1770's and spent some time working on Banks's insect collection. There were many new species in Banks's material and Fabricius described them in his *Systema Entomologica* in 1775. The details given in Fabricius's publications can be linked to the specimens and help establish their provenance. Fabricius also worked on other insect collections in England, including those of William Hunter and Dru Drury. On later visits to London Fabricius made further studies of the collection, including material newly acquired by Banks. For example, the species collected by Nelson on Cook's third voyage were described in the *Mantissa Insectorum* in 1787. The specimens on which Fabricius based his descriptions of the large number of new species in Banks's collections are "types" and have a continuing scientific importance in taxonomy today. Other contemporary entomologists, including Guillaume-Antoine Olivier and Nils Swederus, also worked on Banks's insects and the collection includes species identified or described by them.

The techniques of pinning insects and arranging them in shallow, glass-lidded drawers, familiar today, were well established by the second half of the eighteenth century when Banks was forming his collection. At that time, when a collection was formally arranged, the labels, with the names, but often not other data associated with the specimens, were pinned separately into the drawer bottom. As a result it is not always easy to work out the provenance of individual specimens. As well as the formally arranged material there are about three thousand unsorted specimens in Banks's collection. Both Fabrician and Olivier types have subsequently been found amongst these. Some of the specimens have Banks's or Fabricius's labels, which are sometimes torn rather than neatly cut. These apparently temporary labels often give details, such as locality, which might have been discarded before formal arrangement and would only be available subsequently from the relevant publication. It is also obvious that Fabricius first described the larger and more spectacular species, in 1775, followed by the less spectacular ones, in the *Species Insectorum*, in 1781. So it is assumed that, at least the small and nondescript unsorted specimens, constitute a residue after Fabricius finished his studies.

In *The history of the collections contained in the Natural History Departments of the British Museum*, published at the beginning of this century, Charles Waterhouse records that Banks's insect collection, presumably on receipt from the Linnean Society, was curated in the order given by Fabricius in the *Systema Entomologica* of 1775, although many of the species were described after that date. As noted already the collection was transferred to new cabinets, and Waterhouse recurated it (then or at some later date) to accord with the arrangement published by Fabricius in his *Systema Eleutheratorum* in 1801. In 1863 the named specimens were carefully documented in the museum's accessions register and referred to as "Fabrician types". Waterhouse could not find a number of species that were said by Fabricius to be in Banks's collection. These missing types are also listed in the register. Interestingly, in 1979 some of these missing types, and more than one hundred specimens from the Cook voyages were found in Hunter's collection (now in Glasgow). Possibly Fabricius acted as agent in enabling Hunter to obtain Banks's "duplicates".

Surprisingly, there is no comprehensive record of the collection. Even its size is not known with certainty! Waterhouse, under the acquisitions for 1863, recorded it as 4,081 specimens, but in another place gives a figure of 3,000. The accessions register records 4,690 specimens. Of course many entomologists have consulted and worked on Banks's material since his death. In particular the type specimens have been given attention, but the results of studies are contained in a number of scattered papers. Much work remains to be done and it would be possible to better provenance many of the specimens. For example, in the case of Australian specimens it may be possible to associate particular specimens with the collecting localities given in Banks's diaries and to see if the species occur in the same localities today. There are also interesting questions to be answered about the history of the collection which would throw further light on Banks's entomological interests.

The cabinets currently housing the collection are inadequate and have deteriorated to a point where there is a pressing need to rehouse the material. This will not only assure its continued preservation, but will provide the opportunity to undertake essential conservation work on the specimens. Many of the cabinet drawers have developed large shrinkage cracks which could allow the entry of pests. The drawers are also too shallow and heads clipped from some longer pins are evidence of inadequate curation in the past. Even the clipped pins sometimes need to be inserted at an angle into the drawer bottoms, putting the specimens at risk. Many specimens are mounted low down on short pins. Staging them or remounting on cards, carefully preserving the original pins intact, will allow them to be handled safely and labels to be added. Other specimens are affected by verdigris and careful conservation is needed to ensure their continued survival.

The coincidence of the 250th anniversary of Joseph Banks's birth is apposite: an appropriate time to pursue funding for the essential rehousing and conservation work, and for cataloguing the collection and researching its history. The project will also fill in more details of Banks's life.

S.R. Band, 'Overton: the Derbyshire estate of Sir Joseph Banks', in *Sir Joseph Banks: a global perspective* (eds. R.E.R. Banks, and others), Royal Botanic Gardens, Kew, 1994, 213–216.

20. OVERTON: THE DERBYSHIRE ESTATE OF SIR JOSEPH BANKS

STUART R. BAND

Ashover Road, Littlemoor, Ashover, Derbyshire

The Overton Estate lay wholly within the parish of Ashover. There were other land holdings in the adjoining parishes of Morton and North Wingfield to the east, and another small estate at Snitterton near Matlock, to the south west. However these holdings have not been fully researched and as such do not fall within the scope of this presentation.

Banks first saw Overton in 1762 when his uncle and guardian Robert Banks-Hodgkinson was in command but it was not until his uncle's death in 1792 that he came into full possession of the estate which was to become his working base for almost 20 years in late August and early September as a stage in the long coach journey from Soho Square to Revesby Abbey.

Rich in mineral resources particularly lead, the estate provided a useful revenue mainly derived from the fortunes of the Gregory lead mine in which Banks had a substantial shareholding. As well as its natural resources it also provided a human one in the person of John Allen the practical miner of the "Investigator" voyage. Finally the important contributions made to stratigraphical geology by John Farey at the instigation of Sir Joseph were developed in and around Overton.

Overton was first recorded in a charter of 1293. By 1323 the site was occupied by William le Hunte whose descendants sold the property to Richard Hodgkinson of Northedge in Ashover parish in 1556. The Hodgkinsons then relinquished the property for a hundred years when it was restored to the family following its purchase by George Hodgkinson. His son William was responsible for the rebuilding of the house in its present form between 1693 and 1699.

The large rectangular house was constructed of local gritstone from the nearby outcrop around an earlier core. Signs of this phase are still visible in the east front with its mullioned and transomed windows whereas on the south front of 5 bays and 2½ storeys the windows are sashed. Following William's death in 1732, the house and estate was managed by Robert Banks-Hodgkinson until his death in 1792 when it became the property of Sir Joseph Banks and was used by him as his working base for two or three weeks in the late summer during the annual migration from Soho Square to Revesby. Following the death of Lady Banks, the house and estate was sold and passed through several hands until 1918 when the Clay Cross Company bought the estate for its mineral resources. The house was then let to various tenants until about 1970 when it was sold and became a home for the elderly. More recently the old folk were moved out and since 1990 the house has been standing empty, its future uncertain.

Overton Hall. The house remains substantially unaltered from the time when it was used by Sir Joseph Banks as his Derbyshire base.

Of all the properties occupied by Banks, Overton still remains, superficially much as he would have remembered it on his last visit in 1812.

The purchase of Overton Hall by George Hodgkinson, in the late 1650's set the seal on the development and enlargement of the estate in later years. George Hodgkinson was one of a breed of Derbyshire yeoman who moved into the lead smelting industry during the second half of the seventeenth century. This was enabled by the breakdown of the old smelting élite when the mediaeval methods were superseded by the introduction of new technology in the form of water-powered smelting mills. By the time of his death in 1692 George Hodgkinson was styled "gentleman" in his will. His son William had developed the smelting trade and expanded his interests in a number of directions both at home and abroad and by the beginning of the new century was trading in lead, iron and timber. The extent of his trading interests are revealed by some of his account books which have survived and show that he was exporting large quantities of lead to the Baltic ports, and importing iron and timber.

William's marriage, in 1693, to Elizabeth Ferne of Bonsall united two prominent lead smelting families and produced just one daughter Ann, who was the sole heir to the Hodgkinson fortune. It was the marriage of Ann Hodgkinson and Joseph Banks II, in 1714, which established the link between the two families. Their first child, Joseph III, was born at Overton in 1715, a daughter Letitia Mary followed in 1716 and in 1719 was born their second son William the father of Sir Joseph Banks. Three more children were to follow: Elizabeth 1720, Robert 1722 and Eleanor Margaret in 1723. When William Hodgkinson died in 1732 his will settled the Overton estate on the second son of his daughter Ann providing he changed his name to Hodgkinson. A memorial tablet in Ashover church records the following:-

"William Hodgkinson, of Overton, in this parish, esquire. He was bred a merchant, and added considerably to his parental estate by his industry and frugality, virtues which he practised himself and greatly encouraged in others. His whole estate, improved with an honest and fair character, he left to Mr William Banks Hodgkinson, second son of his only daughter Ann (whom he survived) by Joseph Banks of Revesby Abbey, in the County of Lincoln esquire who in gratitude erected this monument in his memory".

The uncertainty of life in the early eighteenth century removed Joseph Banks III, unmarried, before he could inherit therefore shifting the inheritance one place to the right. William Banks was now heir apparent to the Revesby estate and Robert Banks became Robert Banks-Hodgkinson of Overton. This then was the tortuous route by which Sir Joseph became possessed of Overton following the death of his uncle Robert in 1792.

Lead mining provided the foundation to the fortunes of the estate. By the end of the seventeenth century most of the major veins in the parish had been discovered and worked to some extent with the Hodgkinson family closely involved in their development. The Gregory and Overton mines were the largest undertakings in the parish and during the late eighteenth century, Gregory was the largest in Derbyshire. Banks and his uncle Robert Banks-Hodgkinson held the controlling shares. The rising fortunes of Gregory mine, which reached its zenith and a gross profit of £40,000 between 1770 and 1775, added to the wealth of the estate through the Banks-Hodgkinson family's 12 of the 44 shares. Closure in 1803 effectively brought to an end large-scale mining in the parish of Ashover.

In 1792 Sir Joseph came into full inheritance of the estate following the death of his uncle. One of his first acts as the new incumbent was to instruct George Nuttall, land surveyor of Matlock, to prepare a map of the parish of Ashover and its 9,000 acres. This was to form the nucleus of the surveys of John Farey 15 years later. During the ten days of his visits, a pattern of activity was set which he was to follow thereafter until his last visit in 1812. The days were spent visiting tenants, perhaps in company with his agent William Milnes, or receiving them at the Hall, discussing problems or taking them to task about bad management. On his visit in 1793 Banks made a descent into Overton mine, only a few hundred yards from his front door, to see at first hand the work in progress. In the event this was to be one of his last personal observations of the earth's strata. Age and the onset of gout thereafter restricted his observations to specimens brought by the miners or surveyors such as Geoge Nuttall and John Farey.

Apart from the needs of his tenants, Banks took time to devote to his own interests in the gardens at Overton and an experiment in forestry on 300 acres of moorland of the former common at 900 feet above sea level near Edlestow. Scots Pine were planted in forty foot rows with large square spaces between, which would later be infilled with Larch and other trees. The last visit to Overton was made in 1812 when advancing years and the ravages of gout made travelling difficult.

INDEX